Hybrid Teaching

Pedagogy, People, Politics

Chris Friend, Editor

© 2021 Hybrid Pedagogy Inc.
Chapters © their respective authors
Credits, licenses, and permissions noted on page 181

Published by Hybrid Pedagogy Inc.
https://hybridpedagogy.org
Washington, D.C.

Cover art by Joe Muhrlin
Book design by Chris Friend; derived from Legrand Orange by Mathias Legrand and Vel
Set in Warnock, Hypatia, and Geeza using X₃LATEX

This work is licensed under a Creative Commons Attribution-NonCommercial 4.0 International License, except where otherwise noted. See page 181 for additional details.

First printing, February 2021
ISBN 978-0-578-85235-5

Contents

Foreword — vii
Robin DeRosa

Introduction — xi
Chris Friend

1 Pedagogy

1. **Technology is Not Pedagogy** — 3
   Sean Michael Morris

2. **Building Castles in the Air** — 7
   Stephen R. Barnard

3. **Pedagogy, Prophecy, Disruption** — 13
   Ian Derk

4. **Slow Interdisciplinarity** — 19
   Abby Goode

5. **The Process of Becoming** — 29
   Marisol Brito and Alexander Fink

6. **Learning to Let Go** — 35
   Chris Friend

7. **Seeking Patterns, Making Meaning** — 39
   Sherri Spelic

8. **Messy and Chaotic Learning** — 45
   Martha Fay Burtis

9. **Pedagogical Violence and Language Dominance** — 55
   Maggie Melo

2 People

10. **Trust, Agency, and Learning** — 65
    Jesse Stommel

11. **Confessions of a Subversive Student** — 69
    Leif Nelson

12 **Do You Trust Your Students?** 75
    Amy A. Hasinoff

13 **On Silence** 81
    Audrey Watters

14 **A Soliloquy on Contingency** 85
    Joseph P. Fisher

15 **N=1: Inquiry into Happiness and Academic Labor** 89
    Ioana Literat

16 **From Ph.D. to Poverty** 95
    Tiffany Kraft

17 **When One Class is Not Enough** 99
    Amanda Licastro

18 **Assessing so That People Stop Killing Each Other** 107
    Asao B. Inoue

3 Politics

19 **Pedagogy as Protest** 117
    Jessica Zeller

20 **Critical Citizenship for Critical Times** 121
    Maha Bali / مها بالي

21 **Interrogating the Digital Divide** 125
    Lee Skallerup Bessette

22 **Pedagogy and the Logic of Platforms** 129
    Chris Gilliard

23 **(dis)Owning Tech** 133
    Timothy R. Amidon

24 **Education as Bulwark of Uselessness** 143
    Luca Morini

25 **The Political Power of Play** 151
    Adeline Koh

26 **Ghost Towns of the Public Good** 159
    Pat Lockley

27 **Ed-Tech in a Time of Trump** 163
    Audrey Watters

4 Reference

About the Authors 177
Permissions and Attributions 181
Bibliography 185
Index 203

Foreword

Robin DeRosa

As the calendar year flipped from 2020 to 2021, many of us yearned for respite; here in the United States, with Trump's presidency coming to a close and a COVID-19 vaccine rolling out, there was a sense that maybe—just maybe—things would get better.

But if there's anything we've learned during this past year's dumpster fire, I hope, it was that 2020 didn't start it. The racism and xenophobia that Trump exploited to get elected in 2016 were already there, the nerve waiting to be struck. The financial fallout of COVID that plunged so many people into immediate hunger or homelessness and an undeserving set of billionaires into exponential portfolio growth was a grotesque demonstration of the false promise (or "lie," depending on your mood) of trickle-down economics; Pacific Islanders, Latino, Black and Indigenous Americans all have a COVID-19 death rate of *double or more* that of whites (APM Research Lab Staff, 2021). All of this is to suggest that while Trump and COVID are two viruses that have ravaged America in 2020, the core physiology of our country has long been fertile ground for the fermenting of a deadly national inequity.

And so it happened to be that the date I had set aside to write this foreword was Jan. 6, 2021. As I watched the confederate flag forcefully pushed into the U.S. Capitol and saw one angry white man obliviously wave it in front of the portrait of Charles Sumner, an abolitionist who was beaten on the floor of the senate in 1856 by a pro-slavery congressional colleague, I wondered if now was the time to talk about pedagogy. People, sure. Politics, of course. But *pedagogy*?

One of the core refrains that has echoed since the Capitol siege, and through the last year (especially each time another unarmed Black person was shot dead by a police officer) has been "This is not who we are." Other variations on a theme: "This is not America." "This is not my America." But how do we know who we are? How do we know what America—or any country—is, what it stands for? One possibility is: We learn it.

What you believe the confederate flag stands for has to do with what you learned: at home from the people who raised you, at school from the people who wrote your textbooks and taught you History, and all around you as you consume media (critically, subconsciously, whatever) and interact with your environment. What is learning? Does it require teaching? If so, is teaching always only done by teachers? Is it always only done by *people*? Is learning political? What about teaching?

In the wake of the insurrection at the Capitol, some have again suggested that "The Humanities" are what's needed to combat the deeply intertwined forces of disinformation and white supremacy. Learning history helps us avoid repeating our most shameful past national chapters; studying the ethics that encode our perspectives and engaging in conversation about how our narratives reflect our best and worst human impulses can help us understand the proliferate imaginary that shapes our material world. But others have pointed out that many people at the center of the American white nationalist movement and the fake-news machine have undergraduate and even graduate degrees in the Humanities. Perhaps it's not enough to "know history" in order to assure that history is not repeated. Perhaps "media literacy" is not enough to assure that media is used in ways that don't further encode violence, surveillance, and discrimination into our culture. Perhaps what we need to examine is not just what content we teach (in the Humanities disciplines or anywhere else), but *how we teach it*. This collection stands, in some ways, at the intersection of where the Humanities meets humanity, where we expand outward from the content that delineates our disciplines to the approaches we take to share knowledge and ideas in the hopes of building a world that is more supportive of our collective humanity.

If humanity is (at) the heart of this collection, so too are "politics." Today, on the day Americans watched—many in pride but most in horror—as the symbolic home of our national government was invaded and desecrated, and the process of a free and fair election was literally disrupted, the word seems both frail and complicated. The U.S. Capitol was designed in 1793 by William Thornton, whose wealth came from his family's ownership of a West Indian sugar plantation and who bought slaves even well into his later life as a documented abolitionist. The confederate flag carried into Statuary Hall January 6 was surely meant as a threat against the momentum of Black Lives Matter, but that momentum is integrated deeply with the fabric not only of the Union that defeated the confederacy, but also with the racism that deeply permeates even the righteous symbols of democracy and decency—like the Capitol building itself—that we fantasize run counter to the brutality of the treasonous mobs (2021 Capitol invaders or 1861 southern secessionists). What meaning does a Black Lives Matter sign carry when it is planted in the manicured front yard of a suburban white family who owns a home in a neighborhood where Black people have been systematically redlined out of property ownership? "Politics" is not so much where two opposing poles irreconcilably collide in an epic battle of good vs. evil, but a complex web of interrelated power dynamics that constantly threatens to obscure privilege and culpability.

    I don't mean to suggest that there is any ambiguity about the ethics of certain kinds of oppressive ideologies or events, nor that there aren't real benefits to authentically extended acts of allyship. But I do mean to suggest that it's not enough to know what happened, and not enough to stand for truth, for justice, for the "American way." We also have to radically recenter our collective humanity as we seek to understand or find these things. And one way to do that is to focus on pedagogy.

    People (productively) quibble about the differences between pedagogy and andragogy (and heutagogy, my personal favorite), but each of these words is fundamentally about, to quote Joshua Eyler (2018), "how humans learn." Generally, "education" suggests that humans learn well when they are taught, but the question of what teaching is or should be, and who should do it and how, is certainly an open one. I don't want to pretend I have precise answers here, or even know how to ask exactly the right questions. What I think, though, is that there is learning to do. That this learning can't just be about things, can't just be an absorption of facts or even an illumination of truth. It has to be a journey towards humanity, infusing criticality and creativity and collaboration with a deeper commitment to our common human flourishing. Those who want to facilitate that learning—for ourselves and for others—are the pedagogues. And it is for them—for us—that this collection has been created. I think of this collection as a tool: less an assessment of how things are or should be, and more of an invitation into the messy, ongoing, collective question of how education can/does/should shape who we are and who we will be.

Introduction

Chris Friend

Teaching is an act of radical care. Our teaching influences the students we work with, the institutions we work within, and the communities we live in. It reflects on the past and considers the present to change the future. Teaching that values and supports individuals within networks of learners builds confidence, connection, and collaboration. To put it more bluntly, teaching that values students as individuals helps them to construct their own knowledge and hold agency in their learning.

Letting students create their own view of the world may at first blush sound like neoliberal calls for individualized learning, so it's worth pausing to differentiate these ideas. Today's promise of "individualized" or "personalized" learning is tomorrow's algorithmically determined learning path that buries a student's agency under a mountain of code. That's not what learning looks like, and it's far from personal. When algorithms decide what, how, and when students should learn, those algorithms essentially program the students, as Seymour Papert (1993) observes, and school itself becomes "a machine to perform laid-down procedures" (p. 60). Treating students like predictable, procedural machines deprives them of agency, independence, and the opportunity to learn vital skills related to digital literacy and identity. Students who learn how knowledge builds on observations and critical thinking and reliable information can use their skills to understand the world around them and differentiate truthfulness and deceit.

Edward Snowden (2019), reflecting on how he learned to program by himself after schooling generally failed to challenge him, noted that "a computer would wait forever to receive my command but would process it the very moment I hit Enter, no questions asked. No teacher had ever been so patient, yet so responsive. Nowhere else—certainly not at school, and not even at home—had I ever felt so in control." How much control do students exert over their own learning? Do they let curiosity lead them, or does a program expect them to respond to a series of predetermined prompts? When we teach, we make assertions about control and who should have it. When we dictate what students should and should not learn, we dismiss the significance of discovery. And when we let computers chart a path for students, we imply that the programmers know best how to teach. But programming isn't teaching. It's encoding. And that's not what we want done to students.

Nor is it what we want done to society. In "We're Never Going Back to the 1950s," Derek Thompson (2020) highlights the growth of niche partitioning of our information sources. He points out that "*news*—that is, sources of new information, of varying truthiness" has exploded in recent years, creating what he calls "a phalanx of news publishers" at the same time that "Google and Facebook duopolized digital advertising, creating a situation where publishers were multiplying as advertising declined." Our news has become hyper-targeted out of both intention and economic necessity.

Back in the 1950s, it was generally assumed that all of America gathered around the television to watch evening news on ABC, CBS, or NBC—for those were the only options for same-day news updates. Having only the "big three" media outlets meant it was harder for a single political mindset to take over a large segment of the population. Harder, because each news outlet had to appeal to a broad segment of society. Today, that scenario feels so distant as to sound counter-intuitive. But at the time, the media existed to vet information, process events, and help people understand what was what. The search for truth-with-a-capital-T gave journalists somewhat of a higher calling.

But today, the "democratization" of media has led not to the freedom of information, but to the populism of news sources. Political forces understand that they no longer need to subject themselves to the scrutiny of journalists charged with representing the people and keeping them informed. Instead, political forces now can create their *own* media outlets with their *own* standards for "vetting," thus doing the equivalent with news and information as poisoning a well, ensuring all water coming from the tap is contaminated.

Democratization of the media happened rapidly and generally without the attention, critique, or objection of public or higher education. While others created novel and disturbing ways to capitalize on the attention economy, filter bubbles, and echo chambers, we were off building resource lists and learning modules, teaching one-off classes about Information Fluency with far too little attention paid to ethics beyond copyright rules. (Remember those days?) By focusing on the information, rather than the people, we turned a brewing knowledge crisis into little more than an academic exercise, complete with rote exercises and badges of completion. The one line of defense society has against just this sort of news-media crisis—education— was caught unawares and incapable of protecting itself—or its students, and therefore the public and future society—against the problem.

We need to find new tools to combat the rapid, persistent move toward extreme-conservative presentations of news, media, and truth. And as Kurt Andersen (2020) explains, those who seek to entrench the will of the rich at the expense of an informed public and engaged citizenry have been playing "*such* a long game" (p. 274). We must reorient our approach to hybrid teaching away from technology and toward humanity. We must start seeing hybrid education, blended learning, e-learning, distance education—or any of the other myriad names applied to the combination of a student and technology—as opportunities to build community and care for one another.

Caring for others has never been so vital.

Approach

It's clear by now that threats to a liberating and empowering education are nothing new. It's equally clear that concerns over the public's relative inability to defend itself politically did not start with the latest election, coup, referendum, or summit. Many of the articles included in this book debuted before recent key watershed moments. Yet they all have a relevance to our present moment, suggesting our current conditions have plenty in common with our past. Like Jesse Stommel, Sean Michael Morris, and I (2020a) said in our introduction to *Critical Digital Pedagogy: A Collection*, material in this current volume "feels just as timely now as it did when it was written (and even more prescient)" when revisited after time has passed (p. 1). The concerns of critical pedagogy prove persistently relevant.

Through my work with *Hybrid Pedagogy* over the past five years, I have seen a remarkable (and at times dizzying) variety of approaches to the scholarship of teaching. For some, the scholarship of teaching and learning serves as a chance merely to market themselves and their classroom practices without an attempt to engage and challenge readers. For others, it's an opportunity to share how students have inspired them. And for others, writing about teaching becomes poetic, philosophical, nearly spiritual. I have seen through this variety the many shapes teaching takes and the many ways people come to this work we all share.

This book reflects that variety by avoiding any single point of entry, consistent tone, or uniform approach. The chapters, like their authors, and like our shared pedagogy, promote a more diverse and democratic approach to teaching, learning, and reading. Some authors sound distinctly hopeful and playful—see Brito and Fink, Burtis, Nelson, Morini, and Koh for example. Others challenge us through more somber approaches—see Derk, Goode, Inoue, Bali, and Amidon. Still others allow weariness to stand resolute alongside their optimism—here I think of Morris, Spelic, Melo, Literat, Zeller, and Lockley. Yes, that's a long list of weary authors. But each of those authors presents a heart-felt challenge: We know the work of teaching is difficult; we know our world is challenging; we know the political forces at play work against us. But it is precisely those obstacles which validate the critical importance of our work.

I wish it were in my power to publish a book early in 2021 that could erase the frustrations and hopelessness that spread globally throughout 2020. I wish I could present chapters announcing the fall of black-box algorithms that control the way we shape the minds of future generations. I wish I could point to decisions that protect minority lives and economic futures through the use of classroom technology and critical pedagogy. That, of course, is impossible, and we must continue our unrelenting push forward. So instead, I have included these challenging chapters in the hopes of saying, "I see you. Let's work on this together."

And togetherness, really, is the core of this book. Its three sections—pedagogy, people, and politics—each presume the existence of human connection, that commodity we acutely value after calamity rarifies it. If the pandemic taught us anything, it is the value of human connection and its distinction from technological connectivity. This book gathers together its authors to offer conversational sustenance: wisdom, insight, challenge, and care.

Organization

This volume, in ways serving as part three of a series, extends the work of *An Urgency of Teachers* (Morris & Stommel, 2018) and *Critical Digital Pedagogy: A Collection* (Stommel et al., 2020b) by preserving a focus on critical digital pedagogy but here applying it more directly and deliberately toward efforts of resistance and discomfort.

Opening with pedagogy as Part 1, this book perhaps starts in expected territory. But the twist presented by the first chapter (Sean's claim that "technology is not pedagogy" when many of us have spent months hearing that Zoom is the shape of modern education) foreshadows the unsettling nature of much of the work that follows. Sean draws our attention to what's important—not the tools, not the tech, but the teaching. And by separating the wheat from the chaff, Sean provides necessary clarity to carry us through the book. Part 1 continues with a balance between efforts to make meaning as educators and warnings of the dangers/threats pedagogy can pose to students. It is energized by the work of Sherri Spelic, who in the early days of a president's administration highlighted the importance of meaning-making within a senseless society surrounded by political upheaval. Her observations remain helpful years later, as we continue to struggle with public perception of the media. In the subsequent chapter, Martha Fay Burtis presents her concerns for education's technological shortsightedness with her trademark wit, mixing lighthearted optimism with both wonder and worry. Then, Maggie Melo shares an intimate story of erasure that reasserts the urgency of critical pedagogy.

Maggie's chapter leads us perfectly into Part 2, focusing on people. It begins with a challenge: Jesse Stommel discusses the need for trust in education, mostly by pointing out how elusive it can be. Balancing critique with optimism, Amy A. Hasinoff talks about trust and rebuilds connections among learners of all stripes. The discomfort deepens with a lament over contingent labor, with Fisher, Literat, and Kraft each showing us with beauty and grace—and also a bit of snark—the devastating effect academia has on both its students and its faculty. Then, Amanda Licastro shows how deeply vital compassion is and how difficult it can be to nurture in even the most engaged classes. And finally, Asao B. Inoue closes Part 2 by discussing ways our efforts in the classroom might actually contribute to a less-violent world.

Part 3, on politics, opens with a lament from Jessica Zeller that helps us find our ethical and pedagogical centers and continues the momentum established in earlier chapters. While keeping a solid theoretical grounding, we turn to matters usually seen as less-than-serious. Reminding us of the porous boundaries between classroom and politics, Maha Bali, Lee Skallerup Bessette, and Chris Gilliard show us the intricate connections between socio-political situations and educational policies in stark, concrete terms, with attention to race and legislation taking center stage. Adeline Koh explores the value of play, and Luca Morini looks at forms of learning often labeled "useless"; each in turn sheds light on ways entrenched academics preserve political power. From there, Pat Lockley shares a melancholic lament over bureaucratic decision-making that echoes frustrations from earlier chapters. Pat directs blame toward systemic problems—the same problems then addressed head-on in Audrey Watters' insistent tour de force. Her final chapter leaves us both hyper-aware of problems in today's education system and energized to find better solutions.

Resolution

The need for better solutions to political problems in education becomes critical—in each of the word's meanings—when the entire enterprise is expected to "pivot" online, as it did in early 2020 due to a global health crisis. Using the word "pivot"—which suggests merely the turn of a stationary object—overlooks the sense of direction, movement, and in-process-ness essential to teaching and learning. Classes suddenly required digital tools, home Internet connections, and personal devices to access what previously involved public transportation, lecture halls, and state-subsidized lunches. Suddenly, thoughts of using educational technology went from options to mandates. Nearly overnight, distance education was the only education available. Zoom became ubiquitous. Institutions turned wholesale away from personal connection and exploratory learning toward the use of problematic, dehumanizing delivery models. Calling those changes a "pivot" downplays the disruption they created, both to learning and to living.

Ed-tech often promises pre-packaged solutions and quick fixes to massive challenges—problems like delivering a class online or checking for plagiarism or monitoring attention during testing now have ready-made solutions, we are told, that our institutions can purchase and implement without the involvement—or consent—of teachers and students. These so-called "solutions" strip away the human element from teaching and learning and take control out of the hands of those doing the work of education and place it squarely in the hands (and pocketbooks) of proprietary systems belonging to companies more interested in ownership than in empowerment. We must oppose and resist the widespread, uncritical, high-pressure adoption of educational technology. To make that resistance possible, it helps to have context, perspective, and direction, which together help us better understand the situation's complexity and the shape our response should take.

Addressing complex, dynamic problems requires broad, possibly slow, approaches. When complex problems reach an inflection point, breadth and deliberation can seem irresponsible, adding pressure for rash decisions that sound easy. In this regard, education and politics share challenges as they work to improve people's quality of life, particularly through crisis. And as Barack Obama (2020) observes in a reflection on holding office, "in a crisis people needed a story that made sense of their hardships and spoke to their emotions—a morality tale with clear good guys and bad guys and a plot they could easily follow" (ch. 22). In today's political climate, divisive morality tales have sharpened focus on the perceived line between good and bad. The same can be said of today's educational climate, with a line between progressive educators working to improve student autonomy and agency and cutting-edge technologists working to improve automation and student processing. What, then, happened to progressive educators' "story that made sense of their hardships and spoke to their emotions"? What kinds of stories have we told ourselves?

The stories we tell matter. Obama (2020) adds: "I found myself wondering whether we'd somehow turned a virtue into a vice; whether, trapped in my own high-mindedness, I'd failed to tell the American people a story they could believe in" (ch. 22). As progressive educators, we can too easily get trapped in our collective high-mindedness, becoming our own vices. Indeed, as Donald Macedo notes in his introduction to *Pedagogy of the Oppressed* (Freire, 2014), "educators who misinterpret Freire's notion of dialogical teaching also refuse to link experiences to the politics of culture and critical democracy, thus reducing their pedagogy to a form of middle-class narcissism" (p. 18). In this moment, we must consider the social and political implications of our praxis. We must examine the effects of our work outside our classrooms, both on and offline. Whether education uses technology as a tool of oppression or teachers as a tool for liberation is a conscious choice that we and our institutions must make. Because teaching is always political.

This book was born out of a critical moment in the story of educational technology. That story helps us make sense of how pedagogy, people, and politics influence each other. We see how educators have a responsibility to the people we teach, how education has a responsibility to those who do the teaching, and how teaching itself holds a responsibility to society. That responsibility goes beyond traditional ideas of a civics education or the creation of an "informed citizenry" and into protecting that citizenry. Because as we so vividly saw in 2020, ed-tech, educational institutions, and online platforms will not save us. There's too much inertia carrying us away from agency. Instead, this book aims to unsettle its readers; perhaps some unsettling will do us some good. Because, at the heart of it, education has the power to reinforce or even amplify democracy. By helping students see themselves as empowered members of a larger society, we can do the critical, consciousness-raising work that Paulo Freire (2014) calls us to do. Because, as he says, "There is no such thing as a neutral educational process" (p. 219).

Over the past four years—or four decades, as Andersen (2020) asserts—we have seen the results of unrestrained and calculated greed running rampant across our society and into our education system. But the impulse to dominate others for selfish gain has no place in a democratic education. Indeed, as he reflected on the outcomes of his first term in office, Obama (2020) saw the essential need for:

> government policies that raised living standards and improved education enough to temper humanity's baser impulses. Except now I found myself asking whether those impulses—of violence, greed, corruption, nationalism, racism, and religious intolerance, the all-too-human desire to beat back our own uncertainty and mortality and sense of insignificance by subordinating others— were too strong for any democracy to permanently contain. (ch. 24)

This one book will not "fix" education any more than one head of state can preserve democracy. Nor will this book solve your institution's problems or even show you how to teach an ethical, democratic, inspiring, liberating class next week. But it can challenge thinking, inspire creativity, focus attention, and provide a sense of hope. Hope that, by attending to the challenges of education and the needs of students, we can chart a path forward that allows education to defy surveillance and authority while empowering students to define their world—and then change it.

1 Pedagogy

1. **Technology is Not Pedagogy** — 3
   Sean Michael Morris

2. **Building Castles in the Air** — 7
   Stephen R. Barnard

3. **Pedagogy, Prophecy, Disruption** — 13
   Ian Derk

4. **Slow Interdisciplinarity** — 19
   Abby Goode

5. **The Process of Becoming** — 29
   Marisol Brito and Alexander Fink

6. **Learning to Let Go** — 35
   Chris Friend

7. **Seeking Patterns, Making Meaning** — 39
   Sherri Spelic

8. **Messy and Chaotic Learning** — 45
   Martha Fay Burtis

9. **Pedagogical Violence and Language Dominance** — 55
   Maggie Melo

Technology is Not Pedagogy

Sean Michael Morris

Every year, hurricanes batter the coasts of Florida. But people stay; they don't move away. Every year, wildfires and mudslides endanger those living in Los Angeles. And people stay. Every year, institutions of higher education face budget crises, shortfalls, administrative bloat, and student attrition. And. we. stay. This year, the proverbial shit hit the fan when COVID-19 forced everyone indoors and online. The ensuing rush was a veritable fox hunt for the technological solutions that would provide continuity as we lost our campuses and our communities. Leave the classroom, but get back to class as early as technologically possible. And largely, the "view halloo" was shouted on the first sighting of Zoom and Slack and Flipgrid.

I have been in digital learning in one form or another since 1999. But I have never been asked to speak on the subject more than I have been these last ten weeks. And this is largely due to the fact that my expertise, once seen as fringe or suspect or chancy, has now become the practice upon which education must wager its future.

And yet: My expertise is digital pedagogy—specifically critical digital pedagogy, which resides more in the relationships between teachers and students than it does the delivery of instruction. I'm often thought of as the "tech" guy, but what I actually do is very intentionally human. So as I'm approached with questions about what technologies might help build community online, what platform I might recommend for ensuring students don't cheat, or what digital solution I know of that will enable meaningful discussion, I've found myself answering: Teach *through* the screen, not *to* the screen. Find out where your students are, and make your classroom there, in a multiplicity of places.

When I first started teaching online a dozen years ago or more, my students were scattered across the United States and the Middle East. They were single mothers in rural communities, truck drivers who were rarely in one place for very long, first generation college students without access to a library, and enlisted men and women serving abroad. There was no classroom for them; I had to make one. With words, with conversation, with pictures, with questions. Taking an online class was a risk for many of them. They couldn't be sure they'd finish the term, or that the state funding would come through in time to buy their books. Many of them didn't understand how college could be different from high school, much less how learning online was radically dissimilar from classroom learning.

They came to their screens with little sense of what *there* was there. It wasn't—couldn't be—the technology that created a space for learning.

This crisis facing education didn't need COVID-19. We have been living on an edge for a long time; and to be honest, I'm not sure which way is down. On the one side, there are administrators and administrations that suppose online programs are one solution to the retention of student populations, an answer of higher enrollment for the question of institutional sustainability. These folks have always been much less concerned with the pedagogy of digital teaching and learning than they have been the statistics that reflect success—which in turn mean salability of their programs—and which are supported by very instrumental approaches to education, approaches that Paulo Freire (2014) referred to as "the banking model" of education. Information, or content, is handed to students, and they are then expected to echo back that information in the form of assessments. Rather than knowledge production, these instrumental approaches are focused on knowledge consumption.

But on the other thin side of this edge on which we've been living is a concern about online learning. That it is inadequate. That it's a poor substitute for classroom learning. Among students, online courses are commonly considered easier, and more convenient. And this is because most of the practices of online education assume a universalization of the learning process, one generally founded on behaviorism. In truth, most online practices, courses, and programs are a poor substitute for classroom learning, in part because they attempt to be as much of the classroom as possible. But the only thing that really transfers from the physical to the digital is lecture, rubrics for participation, and, unfortunately, our fear that students will cheat.

We have not *coded* for the human in education, and so, unless we know how to seek it out beyond digital platforms, algorithms, and surveillance tools, the human is largely left out of online learning.

The problem, as I see it, is that no one has started from the beginning. All of the online education industry has jumped the gun. Rather than any single thing bursting onto the scene, there has always had to be a moment of reflection, concentration, contemplation.

What happens when learning goes online?

No educational technology has answered or can answer that question.

The closest I've ever come to hearing an answer was by teaching a three-week course called Learning Online. The title's inversion of "online learning" was intentional, as the goal of the course was, in part, to upset our assumptions about online education. And over those three weeks of digging through the archaeology of our assumptions—me and the students alike, all reflecting, concentrating and contemplating—we came up with several dozen answers, all of them grounded in a humanizing of learning digital.

What's strange about that experience is that the course has been erased by designers of the learning management system (LMS) in which it resided. And so the understandings we all took from our weeks together are all that remain. The technological artifact was unsustainable. Turns out, it was the human experience that persisted.

And it's that human experience we need to not only acknowledge but rely on now.

In "Educational Crises and Ed-Tech: A History", Audrey Watters (2020) reminds us of a statement from Rahm Emanuel, the former chief of staff for President Obama. "You never want a serious crisis to go to waste. And what I mean by that, is an opportunity to do things that you think you could not do before."

In the midst of this crisis, when we are not only faced with abrupt digital teaching and the practices and complications that come with it, but also the inequities of technology upon which a light has suddenly been shone, there are many who want to make uncertainty into opportunity. Corporations dealing in educational technology want us to believe they have the solutions which will not only make this transition to online easier, but will guarantee the success of our students. And advocates for online learning—instructional designers and technologists in some cases, people like me in other cases, and also the stray administrator—see this as a shining moment when everything we know works about online education will come to light. "Teach to the screen," they say. "It's guaranteed to work."

But a crisis is not an opportunity, unless it is for bringing communities together. We can plug our students into the LMS, we can mandate that they turn their cameras on in Zoom, we can use remote proctoring services to ensure they're not cheating on their exams…. But does that constitute teaching? Does that help us develop a sustainable, equitable digital pedagogy?

What happens when learning goes online? This is not a question technology can answer. It's one we need to answer. Teachers, librarians, learning designers, students. Actually good online education comes not from the purchase of another platform, but out of dialogue, out of the will to empower everyone involved in teaching and learning to create together a digital learning that isn't just instrumental, that isn't just performative, but that's authentic, meaningful, and just.

Building Castles in the Air: Pedagogy and the Pursuit of Praxis

Stephen R. Barnard

> If you have built castles in the air...that is where they should be. Now put the foundations under them. —Henry David Thoreau (1995, Conclusion)

There is no one-size-fits-all strategy for teaching with technology, and the decisions about what the right tools are depends as much on the job as it does the laborers. While the challenges posed by the pursuit of praxis-oriented pedagogies may vary greatly depending on educational content and context, we are all affected by the growing mediatization (Hepp & Krotz, 2014) of daily life. The vocational promise of critical digital pedagogy is evident, but how will it be realized? In other words, how do we tone down the hype and get to work realizing the praxis of digital pedagogy?

After spending a week at the inaugural Digital Pedagogy Lab institute in 2015, I find myself inspired by the assortment of amazing people committed to developing critical pedagogies. There were a lot of people sketching out plans to build their own castles in the air, and that planning entailed a lot of tinkering with digital learning tools. Our designs may have differed, but together we worked on the common challenge of how to craft effective pedagogies in today's increasingly networked society.

Digital optimists assert that the tools will make us more democratic, flatten social hierarchies, and make knowledge more accessible and engaging. Meanwhile, digital pessimists worry about the lack of privacy, the substitution of information for knowledge, and the loss of social skills and face-to-face interaction. While neither perspective is wholly (in)correct, both fail to fully explain the opportunities and challenges posed by the digital turn. That is why I (2013) have responded to the "polemics of techno-optimism and techno-pessimism" by making the case for *technorealism*.

Despite notable limitations, and regardless of where you sit on the pessoptimist continuum, the proliferation of digital technologies has the potential to strengthen (or undercut) traditional hierarchies. This potential holds true for a variety of fields—education, politics, journalism, popular culture, etc.—though the dynamics and objectives will vary depending on your position. For education, the time is ripe for pedagogical and institutional innovation. In response to the onslaught of neoliberal pressures for ever-increasing profits, which in turn leads to alienating environments and unremunerated labor relations, we may find solace in the promises of open scholarship and commons-based knowledge. Fortunately, the challenges to traditional classroom hierarchies can also lead to better learning environments, both for the students as well as the teachers.

A praxis-oriented digital pedagogy challenges us to keep our feet on the ground and our heads in the clouds, simultaneously. This means that while we must keep dreaming loftily about the possibilities for *building a better tomorrow*, we must also stay focused on making steps in that direction today. This means selecting the right tools for the job, and being willing to toss them aside when they do not work. Otherwise, we fall into the all-too-common trap of fetishism (Harvey, 2003) and technological solutionism (Morozov, 2013).

2.1 If the Only Tool You Have is a Hammer…

The LMS standard has been to try to be a one-stop-shop for digital learning, but they succeed at being little more than a dumping ground for assignments, grades and readings. It is not because the concept of an LMS is flawed (although it may be), but because the execution of these ideas tends to fall far short of the expectations set by teachers and students. With the widespread accessibility of social media, today's LMSs have big shoes to fill. I have long thought of many LMSs as Web 1.0 tools for the Web 2.0 (or is it 3.0?) era.

Pedagogical tools should be engaging. They should be social and embedded in networks—that is, connected to other digital tools. They should be private by default and public by option. Yet, they should be accessible, and viewable, on all devices. They should be adaptable, so as to accommodate a wide variety of teaching styles. They should primarily be used to augment or supplement—that is, to hybridize—rather than to replace classroom interaction. This suggests that pedagogical tools should be built and selected to perform particular tasks. Too often, decisions to use a tool begs the question of what pedagogical functions they need to serve.

When I began teaching with technology, my primary goal was to *meet students in their world*, to engage them in the practice of critical reflection and discovery. Getting your hands dirty can be fun, especially when you leave room for failure (Wesch, 2015b) and experimentation (Wesch, 2015a). Given that my courses highlight issues of technology and social change, as well as the fact that the vast majority of young Americans (my students included) use some form of social media (Lenhart et al., 2015), a praxis-oriented pedagogy is necessarily a hybrid pedagogy.

In 2011, I began incorporating social media in my classes—starting with blogs, Twitter, Facebook, YouTube, and Tumblr—to make learning more relevant outside the academy. While we were certainly challenged by moments of failure, the majority of my students have found these tools to be a useful supplement to more traditional learning environments. Today, I integrate a variety of digitally mediated assignments, including micro-blogs, online discussion and problem-posing, as well as online research projects. We typically use Twitter to reflect on and discuss course materials as well as to share other relevant information, although the tools can vary greatly depending on the course, student, and project.

These experiences have led to my current, praxis-oriented pedagogy, where I not only teach *with* digital technologies, but also *about* them. Blogging, micro-blogging, and shared documents offer excellent opportunities for collaborative learning. Similarly, digital storytelling and content curation promote creativity and connection. Altogether, this pedagogy goes beyond the expectations of traditional education by helping learners acquire multiple types of capital that are viewed as valuable in today's networked society.

Learners develop digital literacies best when they are acquired in the process of serving other social and intellectual needs. This is yet another reason I have learned to embrace teaching with technology. Critical pedagogy is necessarily dialogical (Barnard & Van Gerven, 2009), and networked technologies offer endless opportunities for engagement—both with the public as well as with fellow students. When learners commit to engaging with networked publics, they find new ways to engage in discovery and serendipitous scholarship, as well as to earn the public recognition that may follow. Given that difficult conversations can turn toxic, and that such toxicity is likely to disproportionately affect already marginalized students, public scholarship may not be the best option for every person, course, or assignment. Nevertheless, when students engage with each other in the network, their sharing and dialogue can lead to shared learning and community building. For example, some of the work my students have done on social media has earned public praise from a variety of audiences, including filmmakers and scholars at other institutions.

I cannot think of a better expression of praxis-oriented pedagogy than an education that emphasizes doing what it seeks to teach. Thus, educators should get to work designing and constructing pedagogical practices that live up to their promises. As Rorabaugh and Stommel (2012) put it, "pedagogy is the place where philosophy and practice meet (aka 'praxis')." Seen through the lens of Thoreau's metaphor, philosophy is the air and practice is the ground. Praxis, then, is the matter that bridges the gap, the foundation upon which our castles must be built. But just as the sky and the ground can feel worlds apart, so too can philosophy and practice. Building bridges that connect (read: synthesize) the two poles is no easy task, but we'll never get there if we don't get to work. Draft up your plans; stop deliberating and start doing.

2.2 DIY Pedagogy and the Pursuit of Praxis: Or, Building Your Own Castle

Pedagogies are like opinions: we all have them, though some are more reflexive, practical and better informed than others. In my experience, the journey toward praxis has been entirely reflexive: as my classroom grew more hybridized, the content also shifted more toward examining the tools we were engaging with. In other words, as the form changed, so did the content. The result has been a pedagogical shift that prioritizes public engagement as well as self-guided, experiential learning.

If the structures we are working with do not engage learners, we must build new ones. But there is no universal blueprint for critical, digital pedagogy. We are drawing up the plans as we go.

Sean Michael Morris (2014) was right to say that "Sisyphus had it easy." Sisyphus was working alone, and his mountain was not nearly as lofty or treacherous as ours. But we are not alone, and we start our journey equipped with all the tools we could possibly need. Although there may be times when we feel isolated, slogging away in institutions of "higher learning" that too often neglect the values at the heart of a liberal education, we must remind ourselves of the community of teacher/scholars working toward shared (and similarly lofty) goals. Higher education wasn't built in a day, nor was it built single-handedly. Like our courses, the key to realizing our potential may not come from within the individual, but from a more even distribution of labor across a wide and diverse network. We have the tools, we have the skills, and we have the community of workers. Now, how about those castles.

Pedagogy, Prophecy, Disruption

Ian Derk

Without consideration of its past, present, or future, critical digital pedagogy may become irrelevant before it begins in earnest. The forces of neoliberalism that critical pedagogues hoped to expose and remove have become quite adept at moving into digital spaces. Online institutions run by for-profit companies attract students from vulnerable populations—the very populations that critical pedagogues aspire to help. For-profit institutions are often a mixed bag, at best, for these students (Maggio & Smith, 2010), but more public and nonprofit institutions model their online offerings to compete with for-profit models. While some professors and academics have resisted changes, the classes they've protected were upper-division seminars rather than developmental or basic courses. Educational experiences that create common ground rather than career or academic tracks have migrated into spaces for efficiency, thus reducing traditional liberal arts and sciences to more closely resemble for-profit colleges' career-focused format.

The rise of the for-profit online classroom is well documented, and the expansion of for-profit education, in part, is the result of decisions made by higher education institutions. While elite institutions were mostly preserved, public schools, especially community colleges, were hurt by the expansion of online education. Spaces for critical, engaged learning in communities gave way to large digital spaces driven by profit motivations. Some of these institutions are starting to falter Kamenetz, 2014, and the space for these failures allow for a critical digital pedagogy to enter online spaces. However, critical digital pedagogues need to consider how they can make critical pedagogy resonate with the public, and use critical theory to examine digital tools and new methods.

But the digital and the critical each face crises of their own. Elizabeth Losh (2012) claims that many people engaged in "hacking the academy" express little interest in the outside world and advocate for open publishing and sharing more out of self-interest rather than shared interest. The argument Losh advances claims that digital humanists fail the call of The Turtlenecked Hairshirt (Bogost, 2010). A lack of engagement with current conditions is also a problem that advocates of critical pedagogy have failed to address in many areas, especially the K-12 education that most Americans encounter (Neumann, 2013). As models and philosophies of personalized learning emerge (Hartley, 2007), with some hopes for co-creation admired by critical pedagogies, these models remain driven by market forces and stand distinct from earlier critical pedagogies of the 1960s and '70s. Models of personalized learning, while changing the dynamic between teacher and student, treat education as another commodity rather than experience.

Critical pedagogy, in some of its forms, focuses on education as the tool to reframe society and challenge market logic. However, critical pedagogy's lack of substantive reaction to the commodification of education has made me question the vitality of critical pedagogy. I discussed with a colleague once whether critical pedagogy was dead. He argued that critical pedagogy is relatively new, and a lack of major results was no indication that critical pedagogy had failed. I would say that critical pedagogy has not done enough in the digital age because critical pedagogues have failed to capture or negotiate against the prophetic ethos.

In *Scientists as Prophets*, Lynda L. Walsh (2013) argues for a particular relationship between a prophetic ethos and scientific research. "Ethos," in her reading, means something akin to "stance" or "role" rather than ethics or credibility. The prophetic ethos is a rhetorical stance with particular motivations, but the most important concept for our purposes is clarification prophecy, or the ability to motivate political decision making. When a deliberative body needed help making a decision, and the future was uncertain, a member of the deliberative body would petition the prophet to access certain (often secret) knowledge that would assist in decision-making. The prophet would deliver a message, but not all prophecies were instructions. Walsh uses Herodotus' description of prophecy and its deliberative interpretation to illustrate the function of prophecy within decision-making. The delegation sent someone to ask for a sign reading. After the first reading signaled nothing but destruction, the delegation asked for a second. The prophet said the city would be saved by a "wooden wall." Some took the sign literally, saying Athens needed to create a wall constructed of wood, while others argued that the prophecy referred to the Athenian navy. Few recorded responses expressed doubt on the prophecy, nor did they claim the prophecy was irrelevant. The delegation wanted certainty and clarification, which the prophecy delivered by focusing the debate on the "wooden wall," but only the body could translate the cryptic words into action. Because, according to L. Walsh (2013), prophets often read signs (an act that balances concealment with revelation), the prophetic ethos might clarify some direction for debate, but it allows for ambiguity while reducing uncertainty for a deliberative body.

Because the prophetic ethos is a rhetorical stance, it is only possible when the prophet speaks to the values of the polity. Expansions into online environments serve desires for convenience, access, and educating a wide variety of people. Some online systems claim to personalize an experience and adapt to students (P. Hill, 2013), thus speaking to a desire for flexible, accessible educational content. These learning-management systems recommend tasks and material for students based on the data of previous students in a manner proponents describe as "similar to Facebook" (P. Hill, 2013). However, some adaptive learning-management systems hide their construction and selection of material without exposing the options or allowing for negotiation. When someone browses other algorithmic systems, like Netflix, there is an option to see the entire catalogue (or at least a great deal of it) by turning off the suggestions. The ability to see all, or at least some, of the possible academic paths in a class would help students retain some agency. Adaptive systems for introductory courses help students with challenging concepts, but deterministic learning in introductory courses could impact a student's future coursework by recommending algorithmically-selected concepts rather than concepts likely to appear in a student's future coursework. Students selected by adaptive systems might enter upper-division coursework with variable skills and preparation, meaning adaptive systems replicate the problems of traditional systems through a less-transparent process.

Beyond the individual course, there are some institutions that plan to create entire degrees with adaptive learning (Kolowich, 2013), perhaps allowing hidden agents to chart the course of a student's entire education. A student who could see the whole catalogue of tutoring and help options—and using these options as supplements rather than instructors (Feldstein, 2013)—could avoid the funneling effect of an adaptive course, but adaptation through hiding choices makes the directive aspects of a learning management system obscure.

Critical pedagogy challenges the idea of inaccessible decision-makers determining the course of education, but algorithmic agents both duplicate and hide the traditional, inaccessible decision-making processes. A hidden algorithm replicates the worst parts of old, cloistered methods of course construction while the statements surrounding the use of algorithms claim to empower the student. The traditional human agent has motives, but the assumption the digital agent lacks motives is wrong, for the algorithm is programmed by a person or organization with motives. Those motives might be unconscious, but they still have consequences when programmed into a system (Tufekci, 2014). Adaptation might work by helping students prioritize or avoid redundant or unnecessary work, but students should know how the system chooses and have the ability to see all the lessons.

Prophets served as types of experts, who were eventually replaced by mechanical experts who use data sets, spreadsheets, and other tools under a cloak of objectivity. These data-driven people made claims that they thought were clear, value-neutral, and simple. However, as L. Walsh (2013) claims,

> Rather than serving as classical evidence in a public debate, the techniques of mechanical objectivity have often proved as legible as a pile of oracle bones on the floor or the ravings of the Pythia; they amount to another confirming sign of the charismatic authority of the science advisers. (92)

Drives toward data sets in educational practices, complex digital scholarship, and esoteric tool development share many similarities to the mechanical expert. Education and digital technology often fall into prophetic spaces because the public knows little about them, and not enough people attempt to interpret those processes for the public and decision-makers. Groups developing digital technology create products that simplify and personalize rather than challenge or examine. The filtering of information limits the ability of people to interpret and understand the world, for they might assume that the information presented is the only information available, rather than an algorithmic guess. If critical digital pedagogues are uncomfortable with the role of prophet, they need to be bold enough explain the connection between technology and work.

Digital technology, to some extent, has disrupted the daily routines of people in personal and professional spheres. The space between someone's private and professional life is shrinking, and maintaining a private sphere divided from work requires an active commitment and employee agency. The syntax and grammar of "work" changes with the addition of technology, and some people enter a post-industrial work environment where work is more efficient because it lacks the constraints of physical space, commuting, or office distractions.

The benefits of post-industrial disruptions have not spread evenly, and thus the inequalities that came during industrialization may remain. And people must remember the industrialized system we see today was not inevitable; the history of scientific management reminds us that technology and practice work together (Lepore, 2009). How people view technology in their lives altered its use, and "efficiency" became beneficial to profit rather than beneficial to all. A critical digital pedagogy needs to start with a critique of disruption, not simply as a critique of itself but to speak to a larger world.

It might be wise to reconsider disruption "as a swear word, in the sense of being potent and rarely used" (N. Schneider, 2014). Disruptions cause damage, and the most vulnerable populations may feel that damage the most. There are lots of discussions about disruption, but precious few interrogate the consequences. The demands for immediate, radical change hurt previous critical pedagogy movements, and critical digital pedagogy should avoid pretending it outsmarted history by changing the terms. Disruption sometimes favors the entrenched and powerful, meaning those wishing to disrupt should plan ahead. A simple archive of interventions, shared in digital spaces, may help.

Earlier critical pedagogies often failed to archive intermediate steps and conflicts. David Tyack and William Tobin (1994) wrote that literature surrounding challenges to the grammar of schools had initial verve and fire, but most literature surrounding those early reforms fell apart. Excitement about the ability to disrupt educational grammar fizzled, and movements were lost. Each person trying to critically reform education had less history, and many people likely made the same mistakes. A long, deep archive of critical pedagogical practices, accessible and shared, might help. There are places that provide decent on-ramps for critical pedagogy (for a fee), but we also need concrete, tangible ideas to create first-order change. While existing technologies have their agendas (Beck, 2014), critical digital pedagogues need to work within them to create change and make contacts. The choices of certain tools may involve uncomfortable compromise, but digital spaces will change before new tools are developed. The deep question for the critical digital pedagogue is, "Can I take the prophetic ethos?"

Critical digital pedagogy is uncomfortable with the prophetic ethos because it allows for abuse, but a critical digital pedagogue must not be afraid to challenge those who offer solutions and obfuscations. Critical digital pedagogues need to take the prophetic ethos because our current system is not ready, but they should demystify critical educational practices and digital technology as well. A stronger network between critical digital pedagogues, decision-makers, and school stakeholders, built over time and carefully, could eventually demystify and change the need for hierarchy. The critical digital pedagogue must help people grapple with uncertainty not by providing direction but by helping them understand structures, but she must speak the language of people. She should change schools with the help of people, not disrupt them for the sake of disruption. Others wishing to disrupt schooling have particular agendas, and they have no problem taking the prophetic ethos.

Slow Interdisciplinarity

Abby Goode

By now, we know that the world is interdisciplinary. We know that, in order to prepare our students for a fluctuating world, we must provide them with opportunities to collaborate across different fields, work in teams to address unsolved, complex problems, and treat them as contributors, rather than just consumers of knowledge. This kind of pedagogy is urgently needed, and in the past few years, scholars such as Joseph Aoun (2017), Cathy N. Davidson (2017a), and Paul Hanstedt (2018) have sounded the alarm. Many institutional communities, such as my own, are working hard to answer these calls for more innovative interdisciplinary curricula and pedagogies. As part of this effort, we envision students collaborating across multiple fields to create their own outward-facing projects or design their own learning experiences.

    Too often, however, these kinds of multi- and interdisciplinary learning experiences only emerge in particular courses (such as First-Year Seminar) or in specific, forward-thinking programs, rather than *across the undergraduate experience*. Too often, students engage in interdisciplinary learning experiences just once, as part of a singular course or project, and then proceed with the business of specialization. Too often, we ask students to collaborate across disciplines without much conversation about what they are doing and why. Too often, the semester ends, and with it, the rich, long-term work of understanding and questioning disciplinary boundaries and norms. Like a major, which involves increasingly complex courses and years of practice, interdisciplinary thinking requires consistent inquiry, exercise, and metacognitive work. Interdisciplinarity takes time.

In our frenzy to be interdisciplinary, we are often quick to label our work and our courses as such, without putting pressure on the implicit knowledge divides and hierarchies that govern intellectual life, without recognizing the vital role that students play in rendering a learning experience interdisciplinary. Interdisciplinarity does not flow from one source. It emerges from a community of thinkers who have, over time, cultivated the desire to learn from fields and perspectives that differ from their own. Given the urgent need to design more meaningful and innovative learning experiences, it might be tempting to briskly hop on the interdisciplinary train without inviting students on board. This fast, frantic version of interdisciplinarity is counterproductive, and it threatens to further destabilize the shifting landscape of higher education. Integrated, project-based, and outward-facing pedagogies are more pressing than ever, but as we welcome students into these forms of learning, let us be thoughtful about our own disciplinary positions and epistemological assumptions. Let us value how students' own backgrounds and ways of knowing enhance, and in fact define the interdisciplinary classroom. This requires slowing down, making space for conversations about disciplinary divides and methods, and recognizing how those divides affect the classroom community.

As we strive to "robot-proof" our students (Aoun, 2017) and prepare them for a "world of flux" (Davidson, 2017a), let us also remember the wisdom of Berg and Seeber (2016) in *The Slow Professor*, which challenges the "frantic pace" of academic life. According to the book's preface, "while slowness has been celebrated in architecture, urban life, and personal relations, it has not yet found its way into education, Yet, if there is one sector of society which should be cultivating deep thought, it is academic teachers" (2016, Page xvii). What would it look like to apply "slowness" to interdisciplinary pedagogies? As a method of teaching and learning, "slow interdisciplinarity" calls us to be mindful, respectful, and curious about each other's disciplinary perspectives—to value ways of knowing that might challenge and enhance our own. This "slowness" does not signify "inefficiency," "smallness," or "ineffectiveness." It is not a plea for more time in developing urgently-needed, integrated curricula and pedagogies for the twenty-first century. And it is most certainly not an obstructionist effort to preserve higher education as it is. In fact, slow interdisciplinarity works against a form of hasty interdisciplinary mania in higher education that can often end up confusing learners and our institutional communities as a whole.

Rather than discuss the dangers of "fast interdisciplinarity," I would like to present a series of lessons that brought me to the concept of "slow interdisciplinarity." I came to this concept last spring, while teaching an experimental, interdisciplinary, project-based course at Plymouth State University (PSU). This course, entitled "American Food Issues: From Fast Food Nation to Farm Stands," asked students to integrate their disciplinary perspectives and work in teams to develop their own initiatives related to contemporary food issues. In this course, we explored how food issues related to consumerism, health and wellness, racial and socioeconomic inequality, and environmentalism. Students identified a challenge related to food in their community and worked to implement their own solutions to issues related to sustainable agriculture, food waste, food security, and food justice. Their work was rigorous, dynamic, and driven by their own interests. Students working on expanding our food pantry, for instance, ended up researching food insecurity in college students. Their project, *not* a pre-selected reading list on the syllabus, brought them to this work. (After all, as Jesse Stommel (2014) writes, "content is co-constructed *as part of* and *not in advance* of the learning.") This is all just to say that this course represents a site, among many others, where "slow interdisciplinarity" can flourish, and it was while teaching this course that I came to recognize the profound complexity of integrated learning, and the continuous, multi-semester efforts that it requires.

4.1 Lesson 1

"Slow interdisciplinarity" entails respect for multiple ways of knowing, recognition of our own disciplinary assumptions and constraints, and willingness to allow other disciplines to impact our learning.

I once taught a course called "Eating American Literature," which focused on literatures of food and agriculture. On paper, the course claimed to be interdisciplinary, but it was, for all intents and purposes, a literature course that highlighted the field's relationship with environmental studies. Nevertheless, to my delight, the class included a range of majors, such as Art, Adventure Education, and Health. Half of the students were English majors and half were from other fields. In this so-called interdisciplinary class, I encountered a major challenge: the tendency for non-English majors to feel alienated and uncertain about their ability to participate in the unspoken-yet-shared assumptions, practices, and ways of knowing that characterized the field of literary studies. English majors, for instance, tend to read *The om-*

nivore's dilemma: The secrets behind what you eat by Michael Pollan (2015) for its figurative language and rhetorical style, while a Biology major might be much more interested in the technical details of botanical descriptions in Pollan (2015), and an Environmental Science major might challenge his characterization of sustainable agriculture practices. (I say "might," because science majors are perfectly capable of reading for figurative language.)

Guess which method of reading *I* favored, and therefore implicitly reinforced in the classroom? When non-English majors contributed to our discussions, it was not uncommon to hear: "Well, I'm not an English major, but…." If students felt the need to preface their remarks in such a way, as if their non-English-major status somehow minimized their contribution, then this classroom was far from an inclusive, integrated environment. This course inspired me to ask: What are the implicit ways in which we reinforce our own disciplinary assumptions and practices in so-called interdisciplinary environments? How do we unconsciously produce and reproduce hierarchies of knowledge and disciplinary divides, even as we label much of what we do as "interdisciplinary"? How can we become more aware of those biases ourselves, signal them to students, and in so doing, welcome other ways of knowing, reading, and thinking about the world into our classrooms?

As an interdisciplinary novice myself, I wanted to avoid unintentionally prioritizing my discipline over others. I wanted to counter the tendency for same-major sub-groups to quickly form and ossify in the classroom. I was eager to see how this might work in an interdisciplinary course on food *that was decidedly not an English class*. The course ended up with the same breakdown of majors as "Eating American Literature": half English majors, half non-English majors. So, for the second day of class, I asked students to respond to the following prompt: "Briefly describe your interests and discipline—the methods, content, and dispositions related to your major field of study—and how they might be useful in a class related to food. *Keep in mind that we have a mix of majors in this class, so imagine you are explaining your major to someone who has no background in the discipline.*" At the beginning of class, I asked students to re-read their disciplinary explanations and capture their major in one sentence for others in the class. I brought chart paper for every major represented in the class, and I invited students to post these sentences at the top of their respective charts: Anthropology, Communication and Media Studies, English, Environmental Science and Policy, Exercise and Sports Physiology, and Psychology. The rest of the chart paper included four quadrants, each with distinct prompts:

1. How does your discipline/field relate to this one? What kinds of similarities do you notice?

2. How might this field be useful for a project related to food?
3. How could you imagine collaborating with someone in this field?
4. What questions do you still have about this field? What are you still wondering about?

I posted these charts around the room. Students circulated and responded to the prompts with respect to fields that were not their own.

My hope was that this introductory exercise would accomplish three things:
1. encourage students to value the various majors and knowledge domains representative in the class,
2. make visible the boundaries and limitations of our own disciplinary perspectives, and
3. spark students' imaginations about how these different disciplinary perspectives might intersect to serve their own projects.

All of these items, especially the third, foster *disciplinary permeability*, or the ability to allow other disciplines—and their respective priorities, methods, and concerns—to impact one's thinking. Examples of disciplinary permeability might include English majors taking up environmental projects that require knowledge of food waste's climatological impact, or Art History majors learning about Criminal Justice as they curate an on-campus exhibition of work by incarcerated artists (Parrish, 2019). To understand that there are other meaningful ways of knowing besides one's own is to engage in what Kathleen Fitzpatrick (2019) calls "generous thinking," and challenge the disciplinary competition and fragmentation that has long characterized the academy. With this exercise, I sought to inspire curiosity about others' majors and intellectual priorities. I sought to break down, at the outset, the invisible borders that divide students from diverse majors, and too often have a deleterious effect on interdisciplinary, project-based courses. And truth be told, I too was curious to learn about other disciplines. I am so glad that we took the time to engage in this exercise, because it set the tone that, in this class, we were going to bring our disparate perspectives together and that these various perspectives were precisely what made the course both challenging and meaningful.

This encouraged students to embrace their own perspectives, while remaining curious about other fields and ways of knowing. As students approached their work in this class, as they designed their own projects in multi-disciplinary groups, they did so with an awareness of their own disciplinary position and how it might differ from those in their group. The projects that emerged, however, were greater than the sum of their disciplinary parts. For instance, one group developed a "Grow Slow" initiative, in which they hosted a "Paint Your Own Plant Pot" session during our campus Earth Day celebration. Working to counteract a widespread culture of fast food and encourage college students to feel more connected to their nourishment, they distributed "blank canvas" pots of soil and microgreen seedlings. With these seeds and pots, they included instructions on how to nurture these seedlings, as well as healthy recipes and information about locally-grown foods. This initiative emerged from the group's collective curiosity about fast food culture. Throughout the semester, they researched the myriad environmental and health impacts of this culture, interviewing faculty members, local business owners, and farmers to learn more about this issue and gain support for their project. This group included a Psychology major, an English Education major, and an Environmental Science and Policy major. Their disciplinary perspectives no doubt influenced their approach to their work. The psychology major, for instance, took an interest in learning about the psychological underpinnings of our industrial eating culture. Ultimately, they developed an environmental education and public outreach project that drew on, but also exceeded their own disciplinary boundaries.

4.2 Lesson 2

"Slow interdisciplinarity" entails a metacognitive awareness of one's own discipline, and an ability to explain that discipline to others.

It just so happens that, during this same exercise, I confronted another challenge to interdisciplinary learning. After students circulated and responded to these questions, we discussed each of these fields, and here is what we discovered: it is incredibly challenging to articulate what we do in our disciplines and why. In fact, some students in the same major disagreed with one another about their field's priorities and methods. Granted, not all majors are alike. Learners cannot be reduced to the stringent codes and norms of their discipline. There were three English Education majors in this class. One was a beekeeper. One was a geographic information system (GIS) expert. One had studied abroad in Japan. Their backgrounds and ways of knowing were far from identical, and delightfully so. Still, this disciplinary exercise, and the challenges that accompanied it, made me wonder. To what extent do we encourage *disciplinary awareness* in our own fields and major courses? To what extent do we prepare students for the kinds of interdisciplinary work that they will encounter in their courses and careers? Do they know what they do in their discipline and why it is important? Can they explain it to someone outside of their field? If disciplinarity happens *before* interdisciplinarity, to what extent are major courses necessary preludes to upper-level, interdisciplinary ones? Do we treat them as such? Can we have disciplinary permeability without a clear sense of disciplinary awareness? Or, if disciplinarity and interdisciplinarity exist in dynamic interrelation with one another—if other disciplines change, enhance, and clarify our own—how can we foster conversation about this relationship within and outside our major courses?

Interdisciplinarity is not only for self-contained programs and courses. It emerges from disciplinary awareness and practice, from multiple metacritical conversations about the boundaries and assumptions of discrete fields. We can jumpstart this process with students by asking them to discuss the significance of their major during the first day of class. For instance, in our introductory course for the English major, we spend the first day of class discussing our preliminary responses to the following question: What does it mean to be an English major? This is highly complicated question, so I ask students to write a longer response to this prompt for the second day of class, and then we spend the entire second day of class debating these responses.

As it turns out, there is more than one way to answer this question. What do English majors *do*, exactly? How? Why? What, if anything, distinguishes the English major from other majors? In upper-level courses in the major, such as Critical Theory, students encounter a similar prompt, but one that builds on these questions: What is the purpose of literary studies and *why does it matter?* The answers to these questions are far from straightforward, and it behooves us to encourage consistent conversation about what we do in our disciplines and why. What do we assume is important? What do we tend to study? What are our implicit values and norms? What is the difference between English and History? English and Media Studies? To what extent do those differences matter? How might we explain the purpose of close reading a poem to a non-English major? The ultimate test of disciplinary awareness is disciplinary ambassadorship, the ability to explain one's discipline to a non-specialist—a necessary skill not only for interdisciplinary capstone courses but also students' post-college lives. How can we cultivate disciplinary ambassadorship throughout the undergraduate experience?

4.3 Lesson 3

"Slow interdisciplinarity" calls us to rethink undergraduate curricula.

Perhaps what prevents us from cultivating this ambassadorship is our current curricular structure, which emphasizes the process of specialization. Students tend to begin their college career with foundational courses such as "Composition" and "First-Year Seminar." Ideally, in their first or second year, students select a major and take introductory courses in that field, alongside general education courses in other disciplines meant to expand their breadth of knowledge. As they take more courses, they progress towards a higher level of rigor and focus within their chosen field, culminating in advanced seminars, internships, and independent research projects. I imagine that this process looks like a pyramid, moving from breadth to increasingly specialized practices and insular communities.

But what comes after specialization in higher education? At what point does this process of specialization open up again, transforming into a practice of sharing, collaborating, and working with people from other fields? What would it look like to rethink the undergraduate experience in a less unidirectional manner, perhaps as a pyramid that re-broadens at the end?

Of course, these shapes oversimplify the complexities of the undergraduate curriculum and student experiences. We could imagine other shapes or perhaps zig-zags that crack open the curriculum (Heidebrink-Bruno, 2014), providing students with multiple opportunities to share their diverse backgrounds and disciplinary orientations with others who might not be in their same field. We could imagine these shapes with less solidly defined borders, taking into account the interplay between coursework, our communities, and our lives. My point here is that the undergraduate curriculum does not have to culminate with disciplinary specialization. "Slow interdisciplinarity" challenges specialized insularity by embracing collaboration across specialities. It involves consistently providing ways for students to step outside of their major, teach their discipline to others, or examine their objects of study from a different disciplinary perspective than their own.

There are small and large ways to do this in our teaching. We can offer linked or federated courses that approach a question or topic from multiple disciplinary lenses. We can meet with other classes and ask students from one class to teach others about a particular disciplinary approach. We can invite students to develop pedagogical materials (Jhangiani, 2017) or contribute to open textbooks (DeRosa, 2018) for subsequent classes. We can read materials by other students in other classes or fields. We can ask students to attend lectures or workshops on our course topics that are outside of our disciplinary perspectives. This semester, in Wilderness Literature, I am inviting students to visit other classes that deal with topics related to the outdoors, nature, and the wilderness from a non-literary perspective. I hope that this kind of activity will give them a broad sense of the other ways of knowing the wilderness, and, ultimately, clarify and deepen their understanding of the literary perspective. These kinds of activities solidify students' disciplinary awareness, but they can also allow them to make their major minor (Davidson, 2017b), preparing them for interdisciplinary projects and collaboration.

4.4 Final Lesson

Slow down; make space.

In an informal course assessment, a student from "American Food Issues" powerfully captured their experience of stepping outside of their major: "Every day when I walk to class, I walk past Boyd [Science Center], and I think to myself, 'I have never taken a course in there and I never will.' There are people I will never have classes with and share ideas with. This class made that possible. It was awesome to connect [with] and learn from English and Science majors. It has made me appreciate the path [that] I chose at PSU." This comment captures the extent to which disciplinary boundaries so deeply shape and divide the world of higher education. But it also gestures towards students' hunger to puncture and learn across those disciplinary boundaries. Importantly, as this comment suggests, interdisciplinary collaboration is not meant to erase disciplinary specialization, rigor, or difference. When implemented slowly, interdisciplinarity fosters an invaluable appreciation and awareness of our chosen "path[s]."

Many leaders in higher education advocate moving away from disciplines, majors, and the pejoratively-termed "silos" altogether, citing their irrelevance in an integrated, collaborative, and ever-changing professional world. But these arguments overlook the ways in which interdisciplinary practice often emerges from effective *disciplinary* practice. They overlook the ways in which interdisciplinary projects often affirm learners' love for their own fields. If we slow down, we can create meaningful flexibility and cross-pollination between disciplinary and interdisciplinary realms. After all, disciplinary specialization is important largely because it prepares students to teach others about their field, deploy their expertise in a range of contexts, and recognize that it belongs to a much wider, varied world of knowledge.

As we develop more interdisciplinary learning experiences, we can benefit from slowing down, consistently asking students to examine their disciplinary assumptions, cultivating metacognitive awareness of disciplinary boundaries, and infusing this pedagogy into the entire undergraduate curriculum (not just particular courses or programs). This requires examining our own assumptions as well. Our training, no matter how sophisticated or inventive, is necessarily constrained by particular norms and priorities. Even in the most specialized or foundational courses, there are always opportunities to make those norms visible—to prepare students to work across disciplinary boundaries.

Often, I hear scholars claim that their work and their field is inherently interdisciplinary. Is that claim, on its own, particularly useful for students, without the intentional process and pedagogy that invites them into that interdisciplinary world? Interdisciplinary pedagogy is not something that magically happens when we rename or revise a course. It does not magically happen when we team teach, link our courses, or list our courses as "interdisciplinary" in the registration catalogue. Like all pedagogies, it "involves recursive, second-order, meta-level work," as Stommel (2014) reminds us. It requires that we constantly alert students to what they are doing and why, and how their own disciplines, interests, and backgrounds relate to others in surprising and delightful ways. Most importantly, interdisciplinarity comes from *the learners themselves*—their fields, their experiences, their ways of knowing. It comes from the questions that they choose to pursue and the collaborations that they undertake. It is a dynamic process, and one that is slower than we think.

The Process of Becoming

Marisol Brito and Alexander Fink

On my luckier days, I am gifted a few invisible moments at pick-up time before my son or one of his preschool classmates calls my name. It's my time to see them as they are without me—a rare opportunity for a parent. Today is a lucky day, and I covertly watch a good friend's daughter balancing in the low branches of a tree. She hesitates for a moment, one last look at the leaves above and the ground below. Her knees bent, lips set in a determined line. Then a slight bounce and she's in the air, arms high, eyes wide, a miniature Amelia Earhart. But even Earhart struggled. The ground is there before she's ready, her surprised feet don't stick the landing and her knees and palms meet the woodchips roughly. There's a short silence before her tears well in time with the pink scrapes on her knees.

    Then her teacher is there, sitting calmly next to her, a hand lightly on her shoulder, "I saw you jump from the fork in the tree. That's a big jump. You landed on both feet, then fell and scraped your knees and hands." He's quiet for a moment, hand still lightly on her shoulder. She nods, tears running down her cheeks, "It looks like you are having some big feelings. Is there something I can do to help you?" She leans into him and he hugs her shoulder. "Sometimes when I fall, I feel surprised, or angry, or embarrassed," he says. They sit together with those words lingering, until she wipes her eyes, jumping up to run to the sandbox.

I've just come from a day of observing classes at a Research One university. I'd seen some impressive things, but I hadn't seen anything like this.

5.1 Praise vs. Encouragement: Beyond 'Good' and 'Right'

"I saw you jump from the fork in the tree. That's a big jump. You landed on both feet, then fell and scraped your knees and hands."

One of our main and often laborious jobs as educators is evaluating students. We empathize with one another over the chore of grading, bidding for the most sympathy, "I have sixty midterms and forty-five final papers and…" We bond darkly over students who, rather than valuing the learning process, take a class, write a paper or complete an assignment "to get an A." But, we should not act surprised by this. For years our students have been rewarded for writing the way teachers want them to with statements like "great work" and "nice job", statements which easily translate to evaluations of the student, rather than evaluations of their work. *Great work* becomes *you are great*. The opposite is devastating: *you are a failure*.

Early childhood education offers a different model, calling on educators to focus on encouragement instead of praise. Popular in the Montessori approach, Praise vs. Encouragement (North American Montessori Center, 2008) avoids evaluative statements and instead strives to use observation to encourage children to be process-oriented rather than product-minded. Though the teacher's statement to the child that falls may sound obvious to us, by using language that focuses on process over product, early childhood educators are able to help students see their work in ways they might not be able to see it themselves, opening opportunities for learning about their own learning (heutagogy). For example, rather than using praising language such as "What a beautiful picture!" a teacher might observe, "I see you used three colors in your picture. Your lines go from the bottom of the page to the top, and some curl in a lot of different directions." This kind of language is helpful in supporting children to nurture an intrinsic rather than external sense of value. Further, if a student adopts the notion that a picture only has value because it is beautiful (or a student has value because she can create a beautiful picture), then it may become difficult for the student to engage in an activity when he or she does not feel confident. The experiential process of drawing a picture becomes—with the language of praise—a chance for success or failure based on perceptions of the end product.

In our own practice in higher ed, it has been difficult to escape from scrawling "good" or "great" down the margins of student papers. After all, we still remember how validating and rewarding such comments felt when coming from a respected professor. However, rather than saying "good" when a student makes a remark we find insightful, we can repeat what we understand the student to be saying, "Ah. So I hear you saying that because the second story gives us more information about Rosa Parks' background and intentions, it more accurately portrays her as a rational agent." If this was indeed the student's intention, this statement recognizes her success in achieving her own goals (rather than ours) while simultaneously giving new words and perspective to the student's writing.

We see using encouragement as a way of supporting what Carol Dweck (2008) calls a "growth" rather than a "fixed" mindset in students. However, we see a lot of challenges to this method. To begin, are students' expectations of "praise" already so deeply embedded in their experience of school that to offer anything different is simply interpreted by them as a failure to completely succeed? Does the phrase "I see that thought connecting to this one" offer something less by way of encouraging further thoughts than "Great idea! Let's connect that to..."? Our plan for the Fall is to talk with our students transparently and to invite them into the process of trying and evaluating this method with us. The process of communicating this change with our students is important because being transparent about the method seems in line with the respect the method intends.

5.2 Care: Learning Beyond the Banking Model

He's quiet for a moment, hand still lightly on her shoulder. She nods, tears running down her cheeks, "It looks like you are having some big feelings. Is there something I can do to help you?" She leans into him and he hugs her shoulder.

In a higher ed classroom, physical touch = potential lawsuit. While we can easily imagine a three or four-year-old leaning into a teacher's embrace, this kind of physical affection immediately draws red flags with older students. However, early childhood education embraces care for many reasons, including building a foundation for appropriate risk-taking that promotes growth. If students know that they will meet care, rather than praise, blame, or indifference, the risk of falling is limited to a few skinned knees—a much lesser consequence than loss of pride, affection or respect.

Though our students may be older, many still fear loss of pride, affection and respect. So, while we may be (rightfully or not) concerned about physical affection, we believe the lesson from early childhood education is of great importance. With this in mind, we have tried to find other ways to "hold" our students. In these ways we hope to separate our care for them as persons from assignment grades or other elements of class. One way to do this is relatively easy (if you don't have a million students): In addition to providing feedback on the "form" of assignments (e.g. "try moving this paragraph", "change from passive voice"), teachers can engage the content of what students write. A response might be, "I'm curious about the direction you took the topic for this paper, does it connect to your interest in education?" This kind of attention to students' personal lives and developing interests might also take place in one-on-one meetings. Curiosity can range from "How are you?" to "I was struck by what you said in class the other day…" All of these demonstrate care and interest that build a foundation for students to risk and fail, knowing that your care about them is not contingent on their success, and further that you will be there to help them back to their feet and encourage them to try again.

We see care as an important, and perhaps often overlooked, element in support of Paulo Freire's call to move beyond a "banking" model of education. Freire critiques pedagogy that "deposits" information in the knowledge bank of a student. While his call has been heard by critical pedagogues, it is still often missed in higher education more broadly. Further, while many models of learning, including progressive active learning practices, more effectively enable students to absorb and utilize information, they do not often help us learn to support students in better engaging risk with courage, shame with vulnerability, ambiguity with leadership, and failure with resilience.

5.3 "Loving them Just as They Are": Treating Students as Human Equals

"Sometimes when I fall, I feel surprised, or angry, or embarrassed," he says.

It's scary to be vulnerable. Perhaps that is in part why, in higher education, students and teachers often collude to construct professors as mythic beings, distant and detached from the everyday risks and falls of human life. When presented with an academic's curriculum vita (the curriculum of a life) we see only a list of accomplishments and successes. This is usually no less true when we stand in front of a classroom. The absence of failures and disappointments makes it difficult for young adults to imagine, and therefore inquire into, the complexity of life. Further, it sweeps away the opportunity to see the way others lean into life, taking risks, and sometimes failing, while simultaneously remaining open and dedicated to growth.

> To be honest, I think emotional accessibility is a shame trigger for researchers and academics. Very early in our training, we are taught that a cool distance and inaccessibility contribute to

prestige, and that if you're too relatable, your credentials will come into question. (B. Brown, 2015, Page 12)

But early childhood educators model vulnerability as a path toward relationships of respect, rather than power. Contemporary educator "Teacher Tom" (Hobson, 2011) calls us to remember, "We are each fully formed, fully valid, fully functional human beings no matter our age." His call echoes the radical premise from which progressive early childhood education greats (e.g. Reggio Emilia and Magda Gerber) orient themselves. We see the respect that follows from this kind of premise in the story above, when the teacher shares his own experience of falling. His words model vulnerability in the face of failure, attentiveness to emotional experience, and the possibility that *in spite of* these feelings, he continues to do things even though he might fall. In this small exchange, he models the idea that we are all peers in life, continually learning and growing, no matter our age or experience.

With these thoughts in mind, we are concerned that some things accepted as standard practice in university teaching make it difficult for instructors and students to experience one another as human equals. For example, the standard lecture model holds that teachers have information, and students receive it—the transfer of information and ideas is a one-way street. Likewise, classes and syllabi are often planned far in advance, minimizing the space for students to influence the direction of the course with their knowledge, experience, and interests. It is almost as if we make a deal with students: we will give you information if you leave your self, who you are and what you otherwise care about, at the door (and we promise we'll do the same). Learning will be better, we imply, if it's not mixed up with actually being people.

To create classrooms that model and practice the ways scholars and professionals engage in work and civic settings with courage, teachers can minimize traditional forms of information transfer (lecture or videos). In their stead, we can create environments where students care about practicing, feel supported in practicing, and are challenged to practice both disciplinary and soft skills. Problem- and challenge- based learning and case-in-point teaching offer strong pedagogical models that re-situate students and professors as co-investigators working learning edges together.

Additionally, we can work to reorient our students' attitudes to one another, as we find our students often struggle with a similar dynamic: feeling they must find ways to impress one another. Inquiring together into this dynamic can begin to genuinely work everyone's learning edges. In some ways, this feels more like coaching than teaching—coaching is not built solely on a history of being a superior player, but rather on using that experience to develop structure and provide resources for other players to improve their skills.

5.4 Conclusions

Trees are scarce in college and university classrooms. But it is not for a lack of trees that students rarely jump and instructors rarely hold them when they fall. At some point along the road to our classrooms, students learn to keep their feet firmly on the ground and as academics we learn that it is detachment, not attachment, that provides a fair and professional environment.

The K-12 system has a lot to gain as universities learn to become more open-access and prioritize public engagement alongside research and teaching. However, universities have allowed scientific attitudes of objectivity and detachment to permeate even the social life and fabric of professors and students. For the most part, people seem to accept this as *the* way things are—there's an unspoken myth that sometime between childhood and adulthood we lose our inclination to experience life, as Max van Manen (2016) writes, as a process of becoming. We forget that what we have accepted as *the* way, is simply *a* way. Van Manen (2016) reminds us that *the* way is always a myth, and suggests that when we open ourselves to a child's world, we see that life, regardless of one's age, is a continuous exercise in possibility. In this way, van Manen (2016) claims, children are our teachers.

> Children are children because they are in the process of becoming. They experience life as possibility. Parents and teachers act pedagogically when they intentionally show possible ways of being for the child. They can do this if they realize that adulthood itself is never a finished project. (van Manen, 2016, Page 11)

We can learn a great deal from those who have practices deeply respecting human beings as beings that are continually in the process of becoming. We believe these educators can help us create classrooms that challenge *the* way and embrace possibility: Classrooms where students jump, and instructors hold them when they fall.

Learning to Let Go: Listening to Students in Discussion

Chris Friend

A class discussion where the teacher pre-determines the outcome is just a lecture in disguise, dressed up to feel student-centered while still being instructor-directed. When a class involves discussion, we owe it to our students to *not* know what's going to happen, lest we start dictating what we want them to think. To truly engage another in a conversation, we respond to the ideas that develop organically; a person who talks without listening delivers a speech, not a discussion. The moment we attempt to set the conclusion of a discussion before it starts, we cheat our students out of an opportunity for honest engagement, and we fool ourselves into thinking we let our students learn things for themselves.

I sensed I had a problem with discussions in Spring 2014, when I taught two consecutive classes that were identical on paper: same course, same content, same classroom. Only the time and the students were different. It took many weeks before I realized how foolish that view was; despite the "on paper" claims, in practice the two classes were nothing alike. What could possibly be more defining of a class than the students involved and the time I spent with them? Yet my efforts to plan and run my classes kept frustrating me—I struggled to keep the classes aligned so that I could remember where we were and what we needed to do next.

Those complaints, which I've heard from many other teachers as we work to simplify our planning, reveal deeply troubling perspectives on how a class operates. It's second nature to talk about how *I* plan and run a class that *I* want to align. It is *I* who does these things. It's as though students aren't a part of that process. We generally believe they don't plan the course, they don't run the course, and that they need to align themselves to the expectations of the course, not the other way around. I've been hearing about and talking about "meeting students where they are" for years, yet here I was, complaining that students, wherever they were, weren't meeting me where *I* thought the class should be.

This semester began with a challenge to my traditional frustrations: I was told pretty late in the summer that one of my 50-minute, 3×-per-week classes would move to 80-minute, 2×-per-week sessions. The rest of my classes stayed M/W/F. I shared with others, including my department chair, that I worried about my ability to keep everything on track. The last time I had classes with different meeting times, I divided the semester's activities in different ways so that the schedules kept up with one another, but I struggled to adhere to that plan. It was forced and imposed, and it benefited no one but my internal need for predictable structure.

My ability to focus was perhaps the greatest casualty. I worried more about sticking to the plan than existing in the moment. Class discussions became an exercise in reaching a goal—a goal I set for what they would do. I devoted more mental attention to where I wanted our conversations to go than I did to what the students were actually saying. I didn't listen fully, with concentration and my entire self. I cheated them out of what they deserved: my attention.

Years ago, I taught at a public high school. Lesson plans were a fact of life...and the bane of mine. I rarely submitted plans to administration on-time. My assistant principal frequently threatened me with memos she called "nasty-grams" demanding compliance with a contractual obligation. On my annual reviews, I could always count on a not-so-awesome report of my "submits lesson plans appropriately" performance. I hated planning. When I wrote a plan for a lesson, I felt I was eliminating the possibility for responsive, flexible teaching. I often wondered how, on Monday, I was to know what students would think about on Friday. The expectation for weekly, by-the-hour lesson plans at the secondary level is the result of a view of education as a predictable, programmatic process that works the same way every year (Strauss, 2014), for every class (Kahn, 2014), and with every student (The Highlander Editorial Board, 2014). Believing we can plan student learning means we aren't listening to them when they arrive.

Sean Michael Morris and I (2013) wrote about the need to really listen to students, saying that "we have an obligation to give them the opportunity to try things." The way I ran classes that semester, I wasn't actually letting students try things in our conversations. Instead, I expected them to say things, and I waited until they said what I expected. It was a farce, and I should have just told them what was on my mind and waited for them to ingest it, old-school style.

In Fall 2014, I tried a different approach. Once I saw that I was falling into the trap of trying to over-plan the semester, I stopped. I refused. I decided that I would set weekly targets, that we would *do* or *make* something each week, but that the way we went about that task would be figured out as a class, in the moment, by paying attention. I still made comparisons from one course section to the next (because old habits die hard, and because humans are pattern-seeking creatures—see chapter 7), but I learned to let go much more effectively than I thought.

Each of my classes had a "class notes" document in Google Drive. Everyone in the class had access to it, and everyone could write in it. I used it as a sort of virtual whiteboard. When preparing for a class conversation, I came up with a couple questions for students to think through. I opened the class notes, wrote the questions there, and went to class. That's it. Then, in class, I told students I'd take notes for them. I wrote while they discussed, talking to one another and not to me. It's a technique I learned from Scott Launier at UCF, who always impressed me with his ability to get freshmen engaged in genuine discussion. So when discussion started, I had my laptop open, with the document projected onto the screen so everyone can see it. Several students had their laptops out, as well, and were inside the document with me. I posed a question and solicited an opening thought. Once the first student started talking, I looked down. I tried not to intervene in the conversation at all, allowing them to shape the dynamic and to determine the ground rules. If there was silence, I looked up. If I then saw hands raised, I'd wave dismissively and say, "I'm not in charge here. You are. Just talk. You figure it out."

This sounds like I'm ignoring and abandoning the students, leaving them to their own survival. But that's the exact point. I leave them to survive the conversation on the merits of their own contributions, not my guidance. I write what I hear everyone saying. I occasionally write a question of my own in the margins. Sometimes students see my questions and respond; sometimes I refer back to them in a conversational lull; sometimes they simply go unanswered. By taking notes, I show I'm listening. By asking the occasional question, I show I'm attentive. By looking at my screen and not at them, I show that I really do want them to be in charge of the conversation.

To me, one of the most convenient side-effects of this approach is that we have notes from the conversation, meaning I can go back to them later and see where we left off. The notes for each class are distinct, following the shape and thinking of the students who were in the room at the time. I don't have to worry about keeping the classes aligned for my sanity, because the content we discussed is recorded for future reference. And the questions I ask students relate directly to what has been said by the students, not what I expect them to say. It's a very honest way of having a class discussion. It's student-centered the way it should be. And now planning for class couldn't be easier.

Seeking Patterns, Making Meaning: Digital Life in the Tangerine Era

Sherri Spelic

How do we as citizens, educators, parents, neighbors and consumers deal with the flood of political messaging in a polarized and polarizing phase in our society's history? Amid the concerns about the crumbling of democratic practices and institutions (Gessen, 2016), the widespread anxiety among individuals and groups on both ends of the political spectrum (Dreher, 2017), how do we maintain our capacity to be critical in our thinking, empathetic in our relationships, and alert in our engagements?

Considering the breathless pace of events accompanied by official and unofficial statements from the U.S. Executive branch, from foreign leaders around the world, from mainstream media, much of which we take in through the constant whirr of social media churn, we run the risk of being buried by the mountains of information we attempt to process and make sense of. Taking time to think about and respond to questions like those mentioned above can seem like untenable luxuries in moments of upheaval. If this is how many of us experience the current political moment, how are our students coping? How do we know (Gold, 2017)?

What I propose here is an approach to information and data sorting which may offer us and our students potentially fresh and unusual ways of seeing the evidence before us while at the same time opening windows into our individual means of pattern-seeking and meaning-making. This is an invitation to a conscious practice of noticing described by John Mason (2002) in his book, *Researching your own practice: The discipline of noticing*. While the book focuses primarily on the act of teaching, the author suggests that noticing can be applied to any existing area of enquiry and is best suited to working on one's own practice. In the introduction, Mason (2002) explains how noticing as a deliberate practice can be applied:

> Every act of teaching depends on noticing: what children are doing, how they respond, evaluating what is being said or done against expectations and criteria, and considering what might be said or done next. It is almost too obvious even to say that *what you do not notice, you cannot act upon; you cannot choose to act if you do not notice an opportunity.*" (p. 7, emphasis mine)

By studying our collections of information and canvassing them for details, we seek out opportunities to know ourselves better and through this process become "more articulate and more precise about reasons for acting" (Mason, 2002, p. 7).

What might you learn about yourself and your habits by sifting through your collections? What's in all that material you've read, shared, commented on, or railed against? These sample questions invite us to conduct an informal inventory and may help reveal our individual patterns of information gathering and organizing. Rather than attempting to track events or political figures, this approach raises questions like:

In response to the election of the 45th president of the United States,
- which memes have you found and liked on social media?
- what are some examples of humor you have liked and shared with others?
- which 3–5 news items featured prominently in your online forums this past week?
- which forms of creative expression have left a significant impression on you?
- whose links are you most likely to open and read?
- what are some things you miss from the time before the election?
- how much news is enough? How much news is too much and in which forms?

In your own writing or commentary,
- which topics have been most prominent?

- which phrases or words have you used most often to describe
    - the U.S. President?
    - his advisors?
    - the current political climate?
- which words or topics do you actively avoid using?

This serves as a sort of starter pack of questions for students and teachers to begin investigating habits and tendencies. Ideally this process has three steps: 1) Selecting one or two questions to focus on, 2) collecting the data, 3) summarizing the findings and drawing conclusions. Most of these questions lean on an assumed degree of social media consumption. We could also ask ourselves how one's take might differ relying on one-way media streams such as television and radio.

What I appreciate about this method is that it encourages us make sense of our pile of data on our own terms, within our own quirky parameters of processing and understanding. And there's plenty of room for fun, surprise and discovery. It's an opportunity to raise questions about which themes draw and hold our precious attention. By looking at *what* we collect we open the door to learn about the *why* of our collecting as well as the meanings we derive from what we find.

Following the election of the 45th president of the United States, I began taking note of the word "stunning" appearing in article headlines or tweets. "Stunning" was often used to describe a certain sense of surprise or alarm on the writer's part. Over days and weeks there seemed to be a visible uptick in this particular phrase and I began documenting examples as they came up.

@docrocktex26 (10/1/16) "Trump's lack of insight and judgment is stunning even for a severely personality disordered person, that's all I'm saying. He's impaired."

@riotwomennn (1/6/17) "An update from @NancyPelosi on intelligence briefing this morning. 'It was really quite a stunning disclosure.'"

@peterdaou (1/14/17) "15. Finally, stunningly, lack of political leadership has led to a methodical series of leaks laying out a prosecutorial case against Trump."

@H_Town_74 (2/4/17) "STUNNING COVER of major German publication. WHOA!! #The Resistance #MuslimBan #CNN #msnbc"

@hudsonvalstrong (2/10/17) "Six stunning aspects of the Flynn scandal"

@MMFlint (2/11/17) "Stunning, massive crowds overflowing at Republican congressmen's Town Hall meetings this weekend across the country. Righteous anger!"

@JoyAnnReid (3/19/17) "This is stunning. Disturbing and completely abnormal. Media orgs at some point are going to have to respond to this stuff."

When I return to those headlines and the referenced articles, many provide significant claims of inappropriateness of various actions on the part of the new president or his inner circle of advisors. Repeated uses of "stunning" indicated to me an ongoing sense of disbelief at the current state of affairs on multiple levels.

My perceived prevalence of the word "stunning" reminds me that many within my particular filter bubble are struggling to navigate this brave new world in which our previous assumptions about fairness, respect, and institutional integrity are being challenged. We find ourselves "stunned" by the wielding of executive power in observably undemocratic ways.

Other matters I noticed when I investigated various aspects of my collections:

- My favored political news and idea sources are overwhelmingly female: @leahmcelrath, @LibyaLiberty, @shreec, @shakestweetz, @nanaslugdiva, @msdayvt, @sarahkendzior. Their links often draw my attention to important and varied aspects of the morass.
- A lot of the humor I like and choose to share requires a certain appreciation of irony.
- I like to call attention to what I see as unique creative expressions in the face of difficult realities.
- Because I live in Austria I have some distance to daily U.S. media output. I use the 7 a.m. 5-minute news broadcast on Austrian national radio to gauge if what I'm seeing late nights on Twitter in terms of outrage and disbelief is considered headline newsworthy at daybreak in Central Europe.

These are just some examples of my preliminary findings. Looking them over, reminding myself of what made me laugh or when my decision to follow a link was rewarded with an insightful read, I can recognize their collective role in providing necessary sustenance and signposts in the disorienting wilderness of events. Wayfinding (Hudson, 2015) is a process relevant not only in our classrooms but in our living rooms, kitchens, and faculty lounges as well. Pattern seeking as a hobby and diversion may bring us closer to what we need to understand about ourselves and each other in troubling times and beyond.

And once we've identified some patterns, then what? I think there are several things we can do. Before we leap into public action, however, perhaps the most essential work we can engage in is the most frequently overlooked: to *sit* and *be* with our patterns. And what I mean by that is carving out a reflective time and space to literally contemplate what we've found. The point is asking: *Who am I* in light of these piles of data I've created or circulated?

For example, in noticing that I now follow and trust more women with regards to resistance reporting and commentary, I acknowledge being fed up with years and years worth of majority white male punditry. Add to that the simple optics of the current White House staff which mirrors that demographic in the most unflattering ways and I become aware of a lack I failed to appreciate previously. Looking at patterns encourages us to register the various filters we are applying.

If we are serious about the *critical* in digital pedagogy, then interrogating our motives in the actions we take is work we dare not side step. We are deeply accustomed to examining outside phenomena in our institutions and systems, reading, assessing, discussing, evaluating and concluding on what we see, hear, and think. Our filters are internal, not always conscious, yet essential to our individual meaning-making processes (Spelic, 2016). Often, we forget that they are there coloring our view, skewing our perspectives.

Sometimes it's easier for us to decide to march or call or retweet than it is to stop and clarify our deeper purpose. One of the traps in observing some of our online behaviors is overlooking the critical aspect of *display*. With my selection of tweets, blog posts, and articles that I share, I create an outward image. What you see and respond to are my *professed* preferences and opinions. Putting things on display does not guarantee their authenticity or honesty, however.

Pattern-seeking nudges us to try to catch our filters on duty; to notice the intricate services they perform. Sitting with our patterns, perhaps silently for a time, just looking at them, we may be able to find beauty in the mess and encounter filters we didn't even realize were switched on.

Messy and Chaotic Learning

Martha Fay Burtis

We in higher education have spent far too long avoiding larger conversations about the Web: what it means to our culture and communities; how it's re-shaping our social and political landscapes; how it's altering the work of our individual disciplines; and, on a whole, what role schools of higher education should be playing in helping our citizenry understand all of these factors.

Many years ago, I was in a meeting with a few like-minded colleagues, all of whom I deeply respected and many of whom were active, vocal members of the open-education movement. When the topic of the Web, digital citizenship, digital fluency, and digital identity came up at the table, it was asked why we weren't dealing with these issues head-on in our curriculums, across our curriculum. And even at that table, with people who deeply understood the issues at hand, I remember a general shrugging of our shoulders, a sense of "Well, what can we do?" and, perhaps more specifically, a surety that our administrations and our faculties, more generally, weren't ready to see a place for these kinds of conversations at the heart of our curriculums and institutions.

But then a few years later, something happened. In November 2016, as everyone knows, in a rather shocking upset the U.S. elected Donald Trump president, and suddenly, almost overnight, the conversation about the Web changed. In the days and weeks following that election, it became clear that, at least to some degree, we owed the election outcome to a kind of structural deficiency of the Web that we had failed to really see. We had been so focused on our own news streams and social feeds that we had failed to see the deeper currents that were running through (and being directed through) our digital spaces. We had failed to see the forest for the trees.

In the aftermath, we find that we live in a "post-truth" world filled with "fake news" and "alternative facts." And all around us, people are pointing at the Web as the engine that allowed all this to advance: It turns out that understanding how search engines work is really important; it turns out that understanding Facebook algorithms really does matter; it turns out that knowing how to create and disseminate information on the Web is a very, very powerful force.

And it turns out that we have a lot of work to do. I want to talk about how we got here, and to do that we need to consider the shape the Web has taken over the last quarter century in our institutions. If we look at the Web we had to begin with, we can basically identify two competing spaces back to the mid to late '90s: The LMS and the tilde space. Let's take a look at each.

8.1 The LMS

The LMS broke big on the scene around 1997–98. Most of the earliest LMSs were actually built at schools, often under the guidance of faculty, and they focused their initial efforts on building systems for delivering content. In fairness that's what the Web was really best at those days: It was a publishing platform, with the magic of hyperlinking built-in. Given those realities, it's not surprising that the earliest LMSs were basically designed to distribute and share content. Eventually, though, vendors began adding other "management" features: grade books, internal email/messaging tools, attendance tracking. As the Web began to evolve, the LMS continued to evolve: Vendors began adding tools for "pedagogy": discussion boards, chat rooms, interactive tests and quizzes, wiki pages and group collaboration spaces.

The forces behind the LMS began to evolve too, turning away from internally managed projects within a particular school or consortium of schools. Instead you saw the emergence of large companies: Blackboard, WebCT, Angel. Their roots may have been in higher education, but their future was in capturing a marketplace, and to do so, they touted a very particular kind of Web environment for teaching and learning. Their spaces were standardized, their features were streamlined. Your students' experiences would also be standardized and streamlined...predictable.

The LMS underscores and codifies a set of beliefs and values: Within our courses we should build standard interfaces; provide standardized features and tools; and promote, among our students, the expectation that their experiences from one course to the next will be standard and predictable.

I have frequent conversations with students who are completely flummoxed when a professor doesn't post their course content, assignments, and grades in our LMSs. If the grade book isn't being used, the students have no idea how to determine what they're earning in class. If assignments aren't posted (with system prompts that text them when they're due), they forget (or think they can ignore) the work. If a reading isn't in the system, rather than ask where they can find it, they assume it's a mistake with the system, and come to class unprepared.

Make no mistake: our students are learning these cues *from us*. Our institutions and the systems we buy are sending them messages about how we do school. They believe they're simply following the straightforward, streamlined rules, procedures, and steps we've told them to follow.

Meanwhile, we're being taught to follow the same set of rules from the other side of the LMS. We are allowing a corporation to deliver a coded set of tools for us to "improve" our pedagogy—whether or not that's what we want to do with our pedagogy—while also turning over our students' work, data, and information to this corporation and its partners. All of this so that our and our students experiences can be streamlined, predictable, straightforward.

8.2 Tilde Spaces

So while the LMS was emerging in the mid to late '90s as an online space for faculty to embrace in their teaching, many universities were also spinning up another kind of space, affectionately called the "tilde space." Schools like the University of Mary Washington (UMW), provided faculty and students with their own folders on a webserver where they could post HTML documents. The name of each user's designated folder started with a tilde character, followed by a person's username. Some faculty did use those spaces for students to publish on the Web. My former colleague at UMW, Jim Groom, has done a lot of writing and thinking about the history of the tilde space. Tilde spaces actually predate the LMS. Jim's done some informal polling around his personal network and heard from faculty who had them as far back as 1993. And with the magic of archive.org's Wayback Machine, we can actually find Mary Washington's directory of student and faculty sites.

In addition to predating the LMS, the tilde space lived squarely within the complexity of the Web—a sort of free-for-all of unhelpful tech where, according to Jim Groom (2014), "users had to create the www directory, change permissions, FTP files, write HTML, etc. In other words, creating and managing a personal webpage on universities servers back in the mid-90s wasn't simple."

Tilde spaces were our schools' first responses to the Web. They were messy, complex, and chaotic. And, they were quickly overtaken because as the Web evolved, our institutions didn't keep up with evolving those spaces. We didn't add scripting or database features, for example, and so the spaces became technologically irrelevant and obsolete. And instead of putting resources and skills into imagining what those spaces could become for teaching and learning, we continued spending more and more money on the LMSs and other services companies that made them offered (or partnered with).

In short, we abdicated our responsibility in higher education. We allowed ourselves to believe that within our schools the Web was easily understood as a commodified, vendor-managed space in which we could just skim along the surface. We could sit in the walled gardens of our LMSs and, with that, believe we were "teaching online." Meanwhile, the Web was evolving into a massive, amazing garden of marvels and monstrosities. Thinking the marvels were merely entertainment and the monstrosities were merely "fringe", we decided we had no greater responsibility to our students, ourselves, and our citizenry than to stay in our walled spaces, posting PDFs and counting discussion board comments.

8.3 Domain of One's Own

At Mary Washington, our first foray into exploring the Web as a space not of predictability, but as a space for possibility, happened in 2004. In August of that year, every member of my department (the Division of Teaching and Learning Technologies or DTLT) got a domain name and open source Web hosting. We went from having one tool in the toolshed (the LMS) to many, many tools from which we could choose.

The tools themselves were open source applications—applications for creating Web sites in many different flavors: blogs, discussion forums, media galleries, wikis, even open source LMSs. In addition, we were working in a space where, if we wanted to, we could learn to build our own tools, or at the very least we could adapt the tools we had.

Suddenly, the Web felt accessible to me in a way it had never been before. I had complete control over a slice of it, and I dove into understanding how it worked. My colleagues and I all started our own blogs. We began experimenting with open source community building platforms as a way to connect our department since, at the time, we all worked in different buildings. We began building custom learning spaces for courses, based on partnerships with faculty.

Working in open source, on the open Web, made possible all the things I had imagined back in the '90s, and it challenged all of those beliefs and values that the LMS underscored. It was possible to build learning environments that empowered students, and not necessarily to the detriment of the course. I could create learning environments in which the interfaces, tools, and features were customized to the needs of the professor and students. And there was simply no reason to assume that the experience from one course to the next needed to be standardized. Open source was infused with a different set of values and beliefs: co-construction, iteration, fast prototyping, extensibility, and, well, openness.

It was probably within three years that we began to ask the question "What if every faculty member and student had this? Their own domain name. Their own Web space to build what they want or need. What would happen and what would change?" Eventually these questions would lead us to a project at UMW called Domain of One's Own (DOOO).

Within our personal Web spaces, the application that captured our imaginations the most was an open source blogging platform called WordPress. WordPress was popular back then; it's even more popular now. Some estimates suggest that it powers close to 30% of the Web. Within WordPress, users can use plugins and themes to extend and alter the platform. Themes let users change the way their site looks; plugins let them change the way it behaves. This extensibility of WordPress is why it has become, for me, a game-changer. When faculty or students have something they want to build on the Web, I can almost always figure out a way they can achieve it with WordPress.

In 2007, we developed a multi-user blogging platform for UMW that was built on WordPress, called UMW Blogs. The system makes use of a special flavor of WordPress called MultiSite which allows us to administer a single core code instance that governs as many individual sites as we need. In other words, we only have to upgrade it in one place. UMW Blogs has been hugely popular for us; in the nine years since it launched, it has had almost 13,000 users and it now contains 11,000 individual WordPress sites. Students have used UMW Blogs to create literary journals, survey properties around Fredericksburg, build online exhibits, connect with the authors of the works their reading, publish their poetry, develop in-depth online resources, and, of course, to blog.

Using UMW Blogs allows us to give any member of the UMW community a WordPress site—really as many WordPress sites as they want. They could activate whatever plugins or themes they wanted (as long as we had made them available), which meant they could built pretty highly individualized sites. We were quickly moving out of the territory of predictability and into a messier, more chaotic space for teaching and learning online. However, within UMW Blogs we still controlled the underlying code. We decided what plugins and themes are available, and we were the ones controlling when upgrades happened. We needed to push ourselves even further.

Today, any UMW student can get a domain name (for the duration of her time at UMW) and open-source Web hosting alongside it. Faculty and staff also have access to the project. Domain of One's Own has some critical properties that I think embody its uniqueness as an exploration of technology in higher education, and I'd like to focus on four of them.

8.3.1 The Naming of Things

The very first step in signing up for Domain of One's Own is choosing a domain name for yourself. We have few limitations on what our faculty and students can choose. We do restrict them to four top-level domains (primarily to ensure that pricing remains consistent): .com, .net, .info. and .org. Beyond that the sky's the limit, but we offer guidelines. In the end our goal is for the naming to represent a moment of taking ownership: a consideration of what a thing is through its naming.

I believe there's something actually metaphysical about the act of naming a thing. I believe that on some level it's the naming that helps call a thing into existence. For many of our students this possibility of creating a space for themselves on the web that they not only can build but that they can actually name represents an opportunity that they've never had before. It certainly represents an opportunity that is nothing like anything else we're asking them to do on the web within the context of their higher education.

As I was thinking about this history, I read an article on CNN about a new kind of cloud that meteorologists have named (Chavez & McKirdy, 2017). It's called the *asperitas* cloud, and it's just been added to the official International Cloud Atlas (yes, that's a wonderful thing I learned exists). The name comes from the Latin word meaning roughness, and the cloud is identified as having "localized waves in the cloud base, either smooth or dappled with smaller features, sometimes descending into sharp points, as if viewing a roughened sea surface from below" (World Meteorological Association, 2017). In the middle of the article, there is this quote from author and meteorologist Gavin Pretor-Pinney:

> When we know the name of something, we began to know it in a different way and when we began to know it, we began to care about it. (qtd. in Chavez & McKirdy, 2017, n.p.)

This idea resonated deeply with me because it captured exactly what it is that I think the naming of a domain represents. In choosing a domain, we hope that students will begin to know their place on the Web differently, and in that knowing, we hope they begin to care, as well.

8.3.2 The Building of Things

In addition to the domain name that we pay for while students are at UMW, we also give them space on an open-source LAMP webserver. In case you're interested, the LAMP part stands for the open source technology stack that sits on the server. That would be Linux, Apache, MySQL, and PHP or Perl or Python as the scripting language.

We choose this stack very deliberately. For one, the open source platform embodies, frankly, openness: The applications that students can install here are all open source meaning that their code is readable and modifiable. We also choose the open-source platform because it is inherently portable for when students graduate. They can take what they build with them, if they so desire.

For the most part what students do in classes that engage with DOOO is build things. They build web sites, primarily using WordPress. But they're not limited to WordPress. We have students in computer science classes who use DOOO as a platform for building their own custom applications. We have students in history classes who install the open-source collection and curation application Omeka. Over the last couple of years we have had a number of students in our digital studies program experimenting with Known: an open source application that lets you distribute your content across various web sites and social networks.

The building of things. The building of websites. This is absolutely the core activity of DOOO at Mary Washington. For many students, they've lived on the Web their entire lives—literally as far back as they can remember they have always been engaged with the Web in some way. I take a survey in my freshman seminar of students' earliest memories of the Web and they regularly are able to recall websites that they used when they were in kindergarten, 1st, 2nd grade. They've lived on the Web their whole lives and yet they have lived primarily in spaces that have been controlled for them by media conglomerates, television networks, schools, and social networking companies. Many of them have never had the opportunity to take back that control and build something from scratch on the Web.

8.3.3 The Breaking of Things

So if you haven't realized by now, there is nothing inherently streamlined or predictable about DOOO. In fact, much like the tilde spaces of the early '90s that Jim described in his blog post, managing a domain is not simple. Users have to grapple with some complicated technical concepts and tasks, particularly when things go wrong. Sometimes it feels like people would like me to tell them that actually we at UMW have developed a foolproof system for ensuring that things never really break, or at least not in any kind of serious ways. Or that we have a tool that we use to fix the really bad breaks, really quickly.

We haven't, and we don't.

The truth is that things go wrong, and they go wrong in all kinds of ways. First, there's the technical kind of "going wrong." Things like installing a plugin that doesn't work with their version of WordPress. The site loads as a blank white page. Or their WordPress upgrade fails. The site starts showing random code instead of a site. Or maybe they upload an image to their site that is too big; it breaks the display of everything so their site becomes unreadable. The list goes on and on. On any given day, some new kind of problem can—and will—arise.

The anxiety that faculty express to me about open online spaces like DOOO is not merely about the technical aspects of the project. They worry about other types of things going wrong. What if students, for example, post things they shouldn't? What if they embarrass themselves, their instructor, the institution? What if they make big, bad mistakes, publicly?

I'm sensitive to these concerns, but I think we need to consider this challenge differently. The bottom line is that there is no keeping our students off the Web. They are on it all the time. I guarantee they are already making mistakes. Who is going to have their back as they figure this all out? Who is going to help them understand when they've made a mistake and how to fix it? I believe we're the ones who have to do this. I think it's our responsibility. We have to forge ahead, despite our fears, and we have to be ready to have those difficult conversations with a student when she overshares, when he says something that could be considered offensive, when they post something they've written that is half-baked and not ready for primetime.

I would rather have that difficult conversation with a student now about a comment they left that seems racist, or a biased news article they shared, or a half-baked idea they espoused. I would rather unpack that with them now. I would rather talk with them, listen to them now. I would rather do all of that now if it means that somewhere down the road, when they're out there in the "real world" they think twice before making an offensive comment, sharing a biased piece, espousing a vile idea, or trusting a false prophet.

8.3.4 The Knowing of Things

I started by talking about ways in which I think we have abdicated our responsibility in higher education to really grapple with the Web as a space to interrogate and interpret, and I want to circle back to that now. Because, in all the talk about DOOO, I think it's easy to get bogged down in the naming, building, and breaking to such a large extent that we begin to see the project purely as one aimed at helping students build a product. And, surely, it is that ability to build something that so many students (and faculty) find particularly compelling and enticing. And, from a practical standpoint, there is something quite wonderful about students graduating from UMW with a rich portfolio of their online work, a digital résumé that they can share with future employers or graduate programs. These products are important.

There are other important aspects of DOOO: WordPress, which I've already mentioned several times and which so many of our students use and learn, is a powerful force on the Web. Because it is used by so many sites, learning it is an actual marketable skill that our students can include on their résumés. This matters, and it's worth pointing out and emphasizing to our communities.

Even more importantly, WordPress can actually serve as an exemplar, a symbol with which our students can grapple as a way towards a deeper understanding. The things they learn to do in WordPress are generalizable to other systems and other online spaces: identifying an audience; honing a voice; organizing and architecting an online space; mixing media to create compelling narratives; considering the interplay between design and content; understanding how Web applications work "under the hood" and how databases and scripts interact; adapting sites to consider accessibility and universal design; connecting disparate online spaces so they relate to each other in synthesized whole; adapting a site as it grows and develops in new directions; responding to comments and finding other spaces and sites upon which to comment; learning how search engines rank sites and how those search engines' algorithms impact the findability of their own site. This list goes on and on, and it leads us to a more fundamental conversation about the Web and its place within our classrooms, our disciplines, and our culture.

8.4 The Web as a Built Space

I've begun to think that we need to push for an approach to the Web that considers it as space that begs of us an interpretive approach. Much like in our specific disciplines we learn how to interpret text, research, data, stories, art, I believe we need to approach the Web as an object of this kind of interrogation and consideration. The Web is not merely the content we read or view. It's not merely the sites we browse or post on.

It is a structured space, coded and built by humans with identities, biases, leanings and agendas. It is an evolving space, one that we will have to always be chasing after in order to understand where it might be headed next. It is a commodified space, in which corporations are determined to make lots and lots of money through advertising, content dissemination, journalism, and digital services. It is a political space in which power and access is not evenly distributed, and where people and groups will always attempt to consolidate and reinforce those power differentials. It is neither an inherently good or bad space, but it can, through its marvels and monstrosities, provide amazing and terrible experiences. It is not streamlined or straightforward or predictable. It is messy, chaotic, wonderful, and awful. This is the Web we need to grapple with, for our students' sakes as well as our own. And there is still so much work we have to do.

Pedagogical Violence and Language Dominance

Maggie Melo

My parents handed the lawyer a plastic binder with a paper insert, "Maggie Melo." Moments before, my parents flipped through the file with approving nods. I wanted to hold the folder too, but instead my parents showed the contents from afar—they didn't want me to touch the pages. A collection of my artwork, awards, and report cards (the good ones at least) sprinkled the sheets. Along with my binder were my older brothers' and younger sister's. The lawyer stacked the folders into the crease of his underarm: "Folks, we have a strong case. You're going to leave the courtroom today as American citizens." I can still see myself seated in the courtroom that day. I was wearing a navy-white floral dress, ivory-hued tights, with my hair tied back with a red ribbon. Not by choice, but by design—they wanted me to epitomize patriotism. After a few years living in the United States, my parents successfully petitioned for American citizenship.

My parents emigrated to the United States in the mid-1980s in search for a better life (their words, not mine). When I was born, my parents decided that they wanted to raise me "American." Raising a Filipino child as an

American meant many things to my parents: It dictated the shows I watched, the games I played, the food I ate, but most importantly the language I spoke. My parents exercised verbal hygiene around us kids. They would speak Tagalog to relatives and friends, but not to us directly. They were explicit in their rationale. My mother noted that she just "didn't want us to get confused." She wanted us to speak *proper* American English. While my narrative isn't unique—there are many communities, nations, and peoples that continue to privilege this variety of English—I continue to have an estranged relationship towards Edited American English (EAE) as woman of color, as a student, and as a teacher in the academy.

In this piece, I meditate on the relationship between pedagogical violence and the teaching of EAE (what my Mom would call "proper English"). I am particularly inspired by the conceptualization of violence Paulo Freire (2014) outlined in *Pedagogy of the Oppressed*, where he draws connections between the oppressor and oppressed to the teacher (oppressor) and student (oppressed) relationship. Freire (2014) notes: "Whether urbane or harsh, cultural invasion is thus always an act of violence against the persons of the invaded culture, who lose their originality or face the threat of losing it" (p. 152). I'm extending the idea of violence to include not only physical pain/and or suffering, but also its application to a person's intellectual, emotional, and social well-being. Although the relationship between violence and teaching is a prominent theme in this piece, at its heart is an argument for centering students' desired learning experiences (Stommel, 2014) in the classroom. I will be invoking the theory of experience from John Dewey (2007) to redirect the sole focus on violence—which, I contend, is inherent to the learning process (hooks, 2014) to certain degrees and is particularly pronounced for students of color (R. L. Allen & Rossatto, 2009)—to student learning experiences in the classroom.

I continue to think deeply about the way EAE has achieved a privileged status in classroom space and beyond. I'm reminded of work from Catherine Prendergast (2008) detailing the enduring pursuit for English proficiency within and outside of the United States. Edited American English is touted as a competence allowing for social mobility and personal well-being for anyone. Specifically, EAE is defined in the Committee on CCCC Language (1974) background statement entitled "Students' Right to Their Own Language" (SRTOL). This statement describes EAE as an English variety that is typically seen in newspapers, magazines, and books; a particular flavor of English that has garnered much attention and prestige particularly within the academy. The SRTOL document continues to be highly referenced, and it has obtained widespread reach to English teachers across the world. The treatise acknowledges the privileging of certain English varieties, such as EAE, and the detrimental downstream impacts favoring specific varieties can impose onto the languages and dialects students bring to the classroom. This call for awareness moves to ensure that students remain agents of their learning and of their composition predilections—even if it's not EAE: The National Council of Teachers of English (1974) affirms "students' rights to their own language—to the dialect that expresses their family and community identity, the dialect that expresses their unique identity." However, I want to further problematize this privileging of EAE by acknowledging the limitations of the invoked "student community."

9.1 The Myth of Standard English Users in the Classroom

The header of this section is a not-so-subtle nod to "The Myth of Linguistic Homogeneity in U.S. College Composition" by Paul Kei Matsuda (2006), in which he reveals a major discrepancy in the composition field. Matsuda (2006) asks: Why aren't all writing instructors, assessment metrics, administration, and research concerned with language difference in the classroom? His questioning is provocative. I'm interested in disrupting this "containment" of the invoked classroom: "Behind any pedagogy is an image of prototypical students—the teacher's imagined audience. This image embodies a set of assumptions about who the students are, where they come from, where they are going, what they already know, what they need to know, and how best to teach them" (Matsuda, 2006, p. 639). He disrupts the roles of composition teachers by revealing a disconnect between their perception of their students and their literacies.

Many teachers, myself included, envision their classroom as monolingual. Matsuda (2006) undoes this perception by detailing the various dialects and languages that many, if not all, students bring to the classroom. Dialects can be defined as a variety of English language used by a group whose linguistic habit patterns both reify and are determined by shared regional, social, or cultural perspectives (SRTOL). This is particularly interesting when considering the way that students with varying literacies are taught or even imagined—how does one justify violence placed upon them, when the "them" is considerably always an unknown from the outset? This is what Matsuda (2006) calls the *myth of linguistic homogeneity*: "the tacit and widespread acceptance of the dominant image of composition students as native speakers of a privileged variety of English" (p. 638). The perpetuation of the myth Matsuda (2006) mentions places students at a disadvantage because students are seen as a contained collective; infringing on a "teacher's ability to recognize and address the presence of differences" (p. 639). With that said, how can the teaching of EAE operate on a meaningful level for individual students—on the plane of experience?

9.2 Students' Right to Their Own Language (Learning) Experiences

For several years, I held a Human Resources Training and Development position where I had the opportunity to facilitate international new-hire orientation to thousands of incoming employees. I remember one day meeting my training class of 200 new-hires from China. I learned that many of them decided to work at the company for varying reasons: to learn about the American culture, to meet new people, or (for a resounding majority) to learn or better their English for professional or academic reasons. Many new-hires, now friends, asked me for off-the-clock help with their English. It was interesting to note the varying "types" of Englishes they wanted to learn. While I conversed in proper EAE intonation and register with one person, another one emphasized their desire to learn slang, to speak casually and colloquially among friends. At the time I didn't think much regarding the distinction between EAE and other English varieties. My main objective, instead, was to honor their chosen learning experience—after all, they each had their own unique set of goals and circumstances driving their desire to learn English. My friend learning EAE was a business major seeking to sharpen her English skills, while my other friend wanted to learn conversational English in order to help him mingle in bars and clubs.

I think about the potential violence I would have imparted onto my Chinese friends and realize that I don't have a sure way to measure or predict it. That is, as their informal English instructor, there wasn't a way to know whether I could be causing my friends harm on a social, political, or even economic level based on the variety of English they wanted to learn; I do know, however, that violence would likely have been imparted onto my friends if I were to teach them an English variety not of their choosing. That is, it would've been incredibly violent to dismiss their agency in deciding their own learning experiences. I took assurance knowing, however, that they were steadfast in their decision to learn a specific English variety. I quickly realized that my biases towards certain Englishes were subordinate to their needs and goals.

As I began to think more critically about the various Englishes I was teaching to my Chinese friends, I was reminded of a piece that I've read by Maha Bali (2015c) where she draws connections between teaching and praxis: "For teaching to be praxis, we need to constantly reflect on what we are doing and why we are doing it and what kinds of effects we are having on the world by the ways we teach and what we do." Maha's work prompted my thoughts on the theory of experience from John Dewey (2007), specifically his advocacy for students' agency in self-identifying which learning experiences are most meaningful for them. Maha and Dewey's work combine to make me realize that teaching English to my Chinese colleagues was less about the variety of English I was teaching, and instead was more about how I could support my colleagues' agency in choosing *their* English learning experience.

Regarding experience, Dewey (2007) critiqued the progressive and traditional school systems with a theory of experience to account for a more meaningful approach towards giving students decision-making agency in their learning experiences. The limitations of traditional and progressive schools influenced Dewey (2007) to develop his theory of experience: "We live from birth to death in a world of persons and things which in large measure is what it is because of what has been done and transmitted from previous human activities" (p. 39). There are two facets to his theory: continuity and interaction. Interaction refers to the experiences a person has from their past, present, and future. Continuity explains the interconnectedness of past, present, and future experiences and how they interact temporally to invite certain future experiences to emerge. These are two fundamental parts of an experience-driven pedagogy that values, first and foremost, the student and their decided learning experiences. Re-engaging conversations on the topic of teaching EAE, I agree with many of the conversations (Perryman-Clarke et al., 2014) advocating for students (Kinloch, 2005) to use their home dialects and languages at school in spite of EAE.

Students should have a right to their own language-learning experiences. That is, although the National Council for the Teaching of English (NCTE) (1974) and González et al. (2006) argue for the preservation of home literacies and languages, I argue in favor of *departing* from home literacies and languages, too. This circles back to ideas of placing student learning at the center of the classroom. Instead of focusing solely on writing assignments advocating for home literacies, code meshing, or code mixing, I believe that the student should make the decision to not only engage one of the options, but to also have the option to depart completely from their home languages and literacies. The disengagement with a home literacy should not be seen as a kind of abandonment or shaming. It instead engages with critical facets of the family, especially immigrant families, on the basis for self-chosen assimilation, survival, and sometimes for needed invisibility. People leave home for a reason, and sometimes that means the leaving behind of their languages. It's a deliberate act of survival.

9.3 Healing and Transparency in the Classroom

I want to offer the idea of transparency (and a couple of its applications) as a way to potentially help lessen pedagogical violence on the front end. In other words, I believe that transparency welcomes disruption of the academy's black box—being forthright with ideas of assessment, the privileging of certain languages, and power dynamics in the classroom. Such topics of conversation cultivate liberatory teaching practices because students are centralized in the learning process. This idea is both inspired and builds from the discussion of "Embracing Subjectivity" by Maha Bali (2015a) where she notes: "We need to stop thinking of external reality as more valuable than subjectivity, to stop treating subjectivity as a barrier to overcome." Embracing subjectivity means the welcoming of critical discussions on biases and power. Demystifying the black box of learning provides students with the opportunity to engage metacognitively: They are more able to situate themselves within the context of the university—how their identity shifts, how it is aggravated, how it can change the way that people treat you, how it can be alienating. In other words, being able to name the violence (potential and past) gives students power. It gives them the ability to move through and against the oppressive structures the academy (and other spaces of course) is built upon.

Transparency in many ways is not a new pedagogical concept; however, I believe it enriches the conversation relating to violence and EAE in a fashion that helps promote experience-based learning. Jesse Stommel (2014) succinctly discerns the differences among teaching, pedagogy, and critical pedagogy: "Teachers teach; pedagogues teach while also actively investigating teaching and learning. Critical pedagogy suggests a specific kind of anticapitalist, liberatory praxis." Within this piece's larger context, a critical-pedagogy approach welcomes various analyses and conversations on difficult topics: notions of social mobility and betterment from higher education and/or how persons have "bought into English," have gained an education, and developed fluency in English, yet are still marginalized as second-rate citizens (Prendergast, 2008). It brings to the forefront the thresholds of material betterment that learning English touts (Shor & Freire, 1987). It speaks to one's own articulation of gender, race, sex, and class as contingencies of upward social mobility.

Being forthright with students allows personal restoration to occur within the classroom writ large. bell hooks (2014) draws from Thich Naht Hanh to conceptualize the teacher as healer for students—a direct acknowledgment of the relationship between teaching and violence. She notes his conceptualization of healing includes the unification of mind, body, and spirit.

Elements of care in the classroom are outlined in Maha Bali's piece where she discerns care on intimate and massive scales (Bali, 2015b), highlighting the need to get to know students individually, to be willing to offer some information of yourself (to challenge the mind-body fragmentation of student and teacher), and to promote holistic well-being of student and teacher.

On a final note, I'm reminded of the importance of scheduling time for healing or, in other words, allotting time to make sense, synthesize, and make meaning of any learning experience. As a teacher and writer, I gravitate towards language to grapple with the feelings and thoughts emerging from learning. For example, when I was an undergraduate, I remember the uneasiness of writing a literacy narrative—a genre that asks students to recount their relationship with writing and reading throughout their lives. The assignment asked how I came to learn "academic English." I was told to include details about my parents, their occupations, and how often they read to me. As an undergraduate woman of color, from a working class family, I felt compelled to perform a certain literacy narrative. I wanted to do well on the project, so I reluctantly opened up about my family. I talked about my parents' emigration to the U.S. in the 1980s. I talked about my mom switching her English on at Carl's Jr. and switching it off at home. I talked about the way my Dad would ask me to talk to others in public, such as the store clerk, because he felt his "accent" made him look silly when he spoke. I wrote about the way our parents didn't speak Tagalog around me and my siblings, and how they didn't read to us so often (working multiple jobs and taking care of children can do that to anyone).

Writing the narrative opened up a part of my upbringing that I hadn't unpacked or given much thought to. The uneasiness of this retrospective analysis was exacerbated when I read and heard about others' literacy narratives. Their well-to-do parents: lawyers, professors, doctors and the like. The in-home libraries and the countless hours of bedtime stories. This assignment, the mere literacy narrative, made evident the countless differences between my colleagues and myself. I wish my professor would've considered the power of language to surface personal challenges.

Language isn't neutral. Writing assignments, too, are framed by ideology. Beyond this text, I'll continue to grapple with ideas of violence and the role I play as a teacher. I'm constantly reminded of the finicky and unpredictable nature of pedagogical violence. Violence doesn't abide temporally. Although a learning experience may be void of violence during one moment, it can still possess the potential to cause suffering in the future. Pedagogical violence is enigmatic at best, yet it continues to move persons away from familiar bonds, knowledges, know-how, and into, perhaps, states of alienation.

People

10 Trust, Agency, and Learning — 65
JESSE STOMMEL

11 Confessions of a Subversive Student — 69
LEIF NELSON

12 Do You Trust Your Students? — 75
AMY A. HASINOFF

13 On Silence — 81
AUDREY WATTERS

14 A Soliloquy on Contingency — 85
JOSEPH P. FISHER

15 N=1: Inquiry into Happiness and Academic Labor — 89
IOANA LITERAT

16 From Ph.D. to Poverty — 95
TIFFANY KRAFT

17 When One Class is Not Enough — 99
AMANDA LICASTRO

18 Assessing so That People Stop Killing Each Other — 107
ASAO B. INOUE

Trust, Agency, and Learning

Jesse Stommel

Technology has the potential to both oppress and liberate. And social media is, right now, rapidly changing the nature of the academic landscape (for teachers, students, writers, and researchers). But there is nothing magical about new technological platforms. We could make similar arguments about Twitter, the Internet, massive open online courses (MOOCs), but also the novel, the pencil, or the chalkboard. I've long said that the chalkboard is the most revolutionary of educational technologies. And it is also a social media. In his 2014 foreword to Paulo Freire's *Pedagogy of the Oppressed*, Richard Shaull writes, "Our advanced technological society is rapidly making objects of most of us and subtly programming us into conformity to the logic of its system. … The paradox is that the same technology that does this to us also creates a new sensitivity to what is happening" (p. 33). So, we feel discomfort when the platforms for or nature of our work change, but that discomfort also causes us to pause and take stock—to interrogate what we do and why we do it.

For this taking stock to happen, educators need to actively guard space for learners and learning. In a continually changing educational landscape, developing trust depends on teachers being advocates more than experts.

Because learning is always a risk. It means, quite literally, opening ourselves to new ideas, new ways of thinking. It means challenging ourselves to engage the world differently. It means taking a leap, which is always done better from a sturdy foundation. This foundation depends on trust—trust that the ground will not give way beneath us, trust for teachers, and trust for our fellow learners in a learning community.

Freire writes, "A revolutionary leadership must accordingly practice cointentional education." And Howard Rheingold (2012) writes in *Net Smart: How to Thrive Online* that participation is "a kind of power that only works if you share it with others" (p. 112).

Connected learning depends, then, not just on agency but also on generosity. In my classrooms (physical, virtual, or some mixture of both), I work extremely hard to keep my own expectations from being the fuel that makes everything go. My only real expectation as a teacher in a learning environment is that students don't look to me for approval but take full ownership of their own learning. And I work to develop trust by showing up as a student myself.

Pedagogical generosity is about making gaps in our work, space for the burgeoning expertise of other scholars and students to fill. It's about advocacy, guarding space for growing expertise, dialogue, discovery, and disobedience.

And bureaucracy is the enemy of learning. In college syllabi, for example, we too often drown students and teachers in policies. Some of these policies are ethical at their core, but every single one becomes an obstacle, if we (teachers, administrators, accreditors, lawmakers) don't trust students to help shape their learning environments. Very little about what happens in a classroom should be fixed in advance. And I mean fixed chairs, inflexible reading lists, predetermined outcomes, and assignments with rules not designed for breaking. It is good to offer guidance and also protections for difference. But, for me, the best outcome for a learning experience is something I never could have anticipated in advance. Trajectories can be mapped, but never at the expense of epiphanies. Unfortunately, our current educational system and its increasing emphasis on standards and mastery, is exactly at odds with this in far too many ways.

Success, then, in digital environments has much less to do with fluency in particular tools and much more to do with our ability to think critically about our tools. I keep getting in trouble on social media for proclaiming my opposition to laptop policies. I'm not actually a wild proponent of laptops or smartphones in the classroom. And I think forcing students to use tech in particular ways is just as problematic as restricting certain uses. For me, it's less about encouraging technology and more about encouraging agency.

(And in the case of allowing laptops, it's often also about acknowledging the needs of disabled students.) Even the assignments I give always have loopholes. I don't believe learning is something that should be policed. Rather, I work to build a learning community through trust—a community in which respect usually comes naturally and is most often arbitrated by the group. There are times I might step in as an "authority," but the situation has to rise well above the clicking of a keyboard or a distracted glance at Facebook. This discussion is not actually about distraction. It's about control. Start by abolishing fixed-seat, face-forward lecture halls where feigned attention is valorized. Then, we can talk about learning and distraction.

Too frequently, suspicion (of both students and teachers) forms the foundation of our institutions and is hard-coded into our technologies.

Concepts like "student data" and "student privacy" are considered far too often in the abstract. We say the words and immediately think of Terms of Service agreements, the Family Educational Rights and Privacy Act (FERPA), or the vague and mysterious cloud. What we need to be thinking about when we say "student data" and "student privacy" are human beings and human relationships, not just legal contracts but also social ones. Abstract notions of hierarchy shouldn't dominate discussions that need also to be about the very real relationships between students and teachers, teachers and administrators, governments and institutions.

When we talk about "student data" and about "student privacy," I think we're actually talking about agency, and I believe real learning is not possible without agency. Agency depends on trust. If we don't feel like the welfare of our data and privacy is in our own hands, we are less likely to feel like full agents in our own learning.

For example, students shouldn't be required by supposedly non-profit educational institutions to publish their theses or dissertations on corporate platforms like ProQuest (see chapter 23). They shouldn't be forced to upload their intellectual property to profit-driven and often predatory sites like Turnitin. They shouldn't be limited from doing public work, asked instead to cloister it inside a walled-garden LMS that controls access to that work. Simply put, students need to be engaged in discussions about data security but allowed to make critical decisions about what happens to their work and where it will live.

In a physical classroom, I'm particularly fond of starting the first class by talking with students about the interface of the classroom—thinking at a meta-level about the effects our environment has on the learning we do within it. We leave no stone unturned in this conversation, talking about how the chairs are arranged, where we each choose to sit, if the room has a "front", whether the windows can open, if the door can be locked, etc. I've watched this activity, or variations of it, scale incredibly well in online classes. I don't want students, myself, or other teachers working inside an LMS, for example, without talking about its affordances and limitations. And the first MOOC I taught (MOOC MOOC, a meta-MOOC about MOOCs) was structured around the idea that none of us can teach or learn freely in an environment without first getting our bearings—without first looking around and thinking about where we are and why we're there. And this is even more important in social learning environments, where we also have to wonder how we're connected—and who isn't there and why. Ultimately, it's this kind of honesty that helps build trust and that helps us build better and more inclusive learning spaces.

Confessions of a Subversive Student

Leif Nelson

I would characterize my lifelong relationship with formal education as a kind of dissonant harmony. As a kid, part of me loved school, yet I would sometimes feel like I was being assimilated by regimented institutions of homogenization. Living "off the grid" with hippie parents was a stark contrast to the school environment of bright fluorescent lights, equidistant rows of desks, and tightly managed schedules. Today, the dissonance continues as I find myself at times upholding and perpetuating systems of conformity in education, while simultaneously trying to disrupt and subvert those systems in order to reveal and dismantle anything that could be inhumane or obsolete. Reconciliation is slow-going and messy. The pendulum swings a wide arc before settling in its consonant center.

In fifth grade, we were given fake money for good grades and good behavior. The "money" was to be spent at "sales" during which our teacher provided trinkets, candy, etc. that we could purchase. This fake economy—supposedly designed to teach us about the "real" economy—was a regulated, glorified reward system for obedience and the timely completion of worksheets. I used the fake currency to set up my own black market. I "hired" classmates to bring toys from home that they could sell, and we expanded our business to other grades that didn't even use the currency except for in my shadow economy. After trying to bribe a second-grader to let me use his basketball at recess, I got in trouble, and my shadow economy collapsed.

That same year, my teacher asked me if she could submit for publication a cartoon I made about saving endangered species. For the first time in my life, I considered the possibility that the things I did in school might have value in the "real world."

In high school, I collected forged hall passes and classroom keys, so I could skip class and spend as much time as possible in the band room writing songs and playing instruments or in the marketing classroom, drinking coffee with friends. Despite my truancies, I genuinely liked thinking and learning, but I preferred to do it on my own time and in my own style. I immersed myself in extra-curricular activities. I was like Jason Schwartzman's character in the Wes Anderson (1998) film, *Rushmore*, founding and presiding over several clubs and organizations. For me, the self-direction and autonomy afforded by extracurricular activities was rewarding in ways that "traditional" coursework was not (e.g., a friend and I founded a DJ club; we were given a sizable budget from the school to buy sound and lighting equipment, and we DJ'ed actual school dances).

I became class Vice President after giving a campaign speech promising to host more punk shows at our school and ending with a shout of, "fight the power!"

A few teachers recognized and responded to my subversive tendencies in interesting ways. An English teacher allowed me to choose and direct a class production of a play. I chose *God* by Woody W. Allen (1975). The teacher essentially stepped aside and let my classmates and me decide how to spend our time in class (we created a fantastic "God-machine" and ultimately performed the play for a special audience of "mature" friends in the cafeteria). A Marketing teacher created a new course offering for me, so I could run the school store and create a newsletter that was distributed to marketing students across the entire state. These experiences were memorable, but they amplified the sense of disengagement I felt in the lecture-based, textbook-centered, breadth-of-coverage curriculum that was the norm in many of my other classes.

In college, classmates would sometimes wonder how I earned passing grades without purchasing textbooks or even attending class (my secret: I showed up on test days). Like with K-12, in college, obedience and short-term memory was too often given disproportionate value. But again, there were exceptions: I took every philosophy course I could. These felt more like book clubs to me—classes were usually unstructured, student-led discussions. But by the time I had earned enough philosophy credits for a major, I discovered that a philosophy degree didn't actually exist at my university. So, I became an English major (finding that these classes had the same type of "book club" format that I enjoyed). Outside of class, I managed the school newspaper and used it as a vehicle to share tips on how to save money in college (e.g., "become a temporary music major to get free guitar lessons"), print cartoons that criticized the financial aid racket, and publish a special issue that contained course evaluation scores for every instructor (which sparked an interesting debate about "open records" law). Outside of class, I found discussions with friends, classmates, and professors to be where much genuine learning took place.

After college, I took a tech writing job at my alma mater's Center for Teaching, Learning, and Technology Development, where I was exposed to constructivist theories of teaching and learning and the technologies that supported them. I was at the ground floor of a boom in online learning in higher education. I remember wishing things like constructivist activities and online learning were more widely adopted when I was in school.

In my work, I came to recognize and name the things that had caused my earlier frustrations with the educational system. The system I had traveled through—with its fake money and multiple choice tests—was still very influenced by behaviorist educational theories. I became drawn to new research trends that explored student-centered approaches. These approaches were often coupled with innovative uses of technology. I had stumbled onto a path that would become my passion.

Today, I am an information-technology director at a university, an adjunct instructor, and a recent graduate of a doctoral program in education. I am about as immersed in the "academy" as one can be, yet I still have this nagging feeling sometimes that something is amiss—that academia as a whole can more effectively develop all individuals to be autonomous and engaged citizens in an unstable world. On one level, this work requires a departure from Aristotlian/Newtonian thinking that has shaped our curriculum (and our organizational structures) into being linear, hierarchical, taxonomical, and essentialistic. It also requires teachers who embrace the values of freedom, empathy, creativity, and inclusivity, and who are permitted to experiment, be reflective, and present their own interests and passions in their teaching without fear of repercussion for dissenting from some status quo. They should also give their students the same latitude.

Paulo Freire (2014) says, in *Pedagogy of the Oppressed*, that

> Education either functions as an instrument which is used to facilitate integration of the younger generation into the logic of

the present system and bring about conformity or it becomes the practice of freedom, the means by which men and women deal critically and creatively with reality and discover how to participate in the transformation of their world.

As I continue to participate in systems and structures that tend to favor conformity, I am also doing what I can to promote critical reflection and practice. I consider whether and how my actions promote values like freedom, empathy, creativity, and inclusivity.

In my professional role, one of the things under my purview is a testing center. When I first visited the center, I was faced with rows of computers, faded white paint, and ominous signs with lists of rules on them. So, the testing center manager and I painted the walls green and hung pictures of nature scenes, citing studies about how nature promotes better concentration and cognition. This was part of a larger goal to shift the emphasis of the center from obedience and punishment to success and support and to shift the ambience of the space from sterile and stress-inducing to natural and relaxing.

Joshua Davis (2013) describes a "radical new teaching method" where a poor, rural school in Mexico produced some of the highest test scores in the country. The teacher who is profiled in the article said he was inspired by Sugata Mitra and the hands-off approach of asking questions rather than providing answers. This rethinking the role of a teacher from one who simply presents facts to one who asks questions in a subtle way challenges the assumptions of authority in student-teacher relationships.

As a teacher, I try to ask good questions and admit my own limitations as an authority in a world where content and information is always evolving. I encourage students to draw their own conclusions, have informed opinions, and use good filters to critically question anyone (or anything) purporting to know some infallible truth.

As a lifelong learner, I still question everything. Despite my skepticism, I need to reconcile myself to the fact that the majority of my life has been connected to educational institutions and academic environments in some way. Educational institutions are not perfect, but it is because I love education and think it is one of the most important human activities that I always look for things that can be challenged, reimagined, or improved. Educational institutions are made up of individual people making decisions and doing work. Educational institutions can be self-perpetuating machines, steeped in traditions and unquestioned ideologies. Occasionally the gears need to be jammed up so basic assumptions can be (re)examined. Both educators and administrators should see creativity as a boon rather than a burden (Westby

& Dawson, 1995). And they should help all students flourish by asking good questions, providing positive support and encouragement, and sometimes—despite impulses to coerce, incent, and control—they should get out of the way and let students lead.

Do You Trust Your Students?

Amy A. Hasinoff

When I began teaching, I focused on content and rigor. I made the rookie mistake of designing my first course in ways that would have worked perfectly for me as a student. The problem was that prep school had made me into a different kind of college student than the ones in front of me who mostly came from underfunded public schools, returned after years away, worked full time, and supported families.

When my initial approach didn't go so well, I was annoyed that students were not respecting my authority (Singham, 2005). I was exasperated by students who plagiarized, came to class unprepared, or showed up late. I was constantly frustrated by the few students in the back row every semester who talked over me or over other students who were contributing to a discussion or nervously giving a formal presentation. I was especially bothered by being mistaken for a student or addressed as "Miss."

I responded with more rules, harsher penalties, weekly reading quizzes, and detailed rubrics. This basically worked because I was using grades as leverage to get the results I wanted. The students learned a lot, but I was frustrated with the feeling that my job was becoming more about explaining and enforcing rules than about teaching and learning.

When I started teaching online, I took a similar approach. I needed a way to ensure they'd do the work even though there was no lecture or class to attend, so each week students were required to take quizzes, answer an ever-growing list of study questions, and post notes about the readings. Each of these things needed to be graded individually.

What that meant is that I spent most of my time on student feedback for that course explaining to students where they'd lost points on each assignment. But I noticed that students who misunderstood a concept in the study questions often hadn't improved when it came up again on the final exam. Turns out, when we give students feedback and a grade, they often pay more attention to the grade (Butler & Nisan, 1986), and may not even read our carefully written comments.

All of these things that frustrated me about teaching led me to look for alternatives. I found critical pedagogy (hooks, 2014) and learner-centered teaching (Doyle, 2012). I was fascinated by studies showing that grades are often ineffective, arbitrary, and demotivating to students (Schinske & Tanner, 2014). I was convinced by Stuart Tannock (2017) that grading might undermine one of public education's key goals of fostering "critical, reflexive, independent and democratically minded thinkers" (p. 1345). And I was persuaded by arguments from Alfie Kohn (2011) and Jesse Stommel (2017) against grading and work from Vicky Reitenauer (2017) on the benefits of self-grading.

So I started experimenting (Hasinoff, 2017b).

With my first attempt, I asked students to reflect on their work and calculate their own points each week, including figuring out late penalties for themselves. I gave them very detailed and complex guidelines for these weekly self-assessments (Hasinoff, 2017a) that ran four pages long. When they made mistakes applying the policies to themselves, I corrected them.

Around halfway through the semester, I realized that I hadn't fundamentally shifted my mindset about grades or power. I'd just outsourced the process of applying my strict rules to the students themselves. While most of them really enjoyed having even this small amount of control, I was still spending a lot of time explaining and enforcing policies for the students who hadn't understood my system. Why had I set it up this way?

I remembered a statement I heard Jesse Stommel make at Digital Pedagogy Lab in Vancouver last summer. As one of my track leaders, he began by introducing himself and describing his approach to teaching: "Start by trusting students."

That was my problem: I didn't feel like I could trust my students. Instead of having empathy for them (Friend et al., 2015), I realized I'd been holding a bit of a grudge against students (Bayers & Camfield, 2018). I had been entering classrooms anticipating all the problems and incivilities (Boice, 1996) I had seen before. I found it even harder to trust students in my online courses, where I usually can't read tone or body language, and there's little opportunity for the casual interactions before or after class that help build a relationship over time.

I also felt like I couldn't trust students when it seemed like they didn't trust me and my classroom authority. For example, in my first semester teaching online, I created weekly short quizzes which added up to a total of less than 5% of their final grade. Because the purpose was to help students check whether they'd understood the course materials, I chose the "unlimited attempts" setting for the quizzes. A number of weeks into the semester, I looked at the data and saw that a couple students were making 10 or even 20 attempts on the quizzes in rapid succession until they got to 100%. While I didn't have a policy against this, I had explained the purpose of the quizzes so I felt like these students were gaming the system and disrespecting the learning environment I tried to create. At the time, instead of approaching those few students individually, I figured that none of them could be trusted, and I changed my policy to limit each quiz to 2 attempts.

The more I learn about critical pedagogy, the more I realize that starting with trust is vital. In chapter 10, Stommel describes the importance and value of trust in education:

> Learning is always a risk. It means, quite literally, opening ourselves to new ideas, new ways of thinking. It means challenging ourselves to engage the world differently. It means taking a leap, which is always done better from a sturdy foundation. This foundation depends on trust—trust that the ground will not give way beneath us, trust for teachers, and trust for our fellow learners in a learning community.
>
> ...
>
> And bureaucracy is the enemy of learning. In college syllabi, for example, we too often drown students and teachers in policies. Some of these policies are ethical at their core, but every single one becomes an obstacle, if we (teachers, administrators, accreditors, lawmakers) don't trust students to help shape their learning environments.

When I first started teaching, I would have scoffed at these ideas. Rules and harsh policies seemed like a bulwark against the vulnerability I felt as a young female professor. Bureaucracy felt like a safety net. Rubrics and grades seemed to provide fairness, clarity, and control. This is why my first experiment with self-assessment still included a lot of rules.

Next semester, I'm going to change the way students do their weekly self-assessments in my online course. Instead of calculating their points each week or grading themselves, students will set their own goals at the start of the semester and then reflect each week on whether and how they've met those goals. I'll respond to them with comments and suggestions. Then, only halfway through the semester and at the end, students will propose a letter grade for themselves and provide evidence to justify it.

But what if I disagree with a student? Will I just trust them to determine their own grade?

In *Hacking Assessment*, Starr Sackstein (2015) says students should conference with instructors to resolve disagreements about their self-assessed grades and argues that it's important to let students have the final say. For small discrepancies, I think it'll be easy to allow the student to decide—if it seems to me like they've earned a B and the student is certain it's a B+, honestly, what's the difference? Tons of research shows that grades can be pretty arbitrary anyways (Schinske & Tanner, 2014).

On the other hand, what if I think they've earned a C, but they think they deserve an A? I've heard from others who use self-assessment that they have simply never come across this issue. But if it happens, and if I'm really going to "start by trusting students," that means I won't just unilaterally decide to record a C on their transcript. Instead, I'll have a conversation with the student and figure out where they're coming from and why our perceptions of the work differ so much. As long as we're both acting in good faith, we should be able to hash it out and agree on a final letter grade.

And that good faith depends on a relationship of trust with the student built over the course of the semester. Without trust and an understanding of the rationale, self-assessment is more likely to feel uncomfortable, awkward, and like it's useless busy work. One student wrote last semester: "I think it is somewhat pointless to self-assess if ultimately the teacher is giving the grades." This was indeed a good point, because in my first version of self-assessment, I sometimes overruled students who had misunderstood my complicated self-grading guidelines.

But they're also right in a larger sense—no matter what structures we set up to try to give students freedom, they may still experience the classroom as a site of control and domination. After all, the instructor still determines the student's final grade on their transcript.

When I first read *Teaching to Transgress* (hooks, 2014) my initial reaction echoed my student's skepticism: Given the fundamental power relationship between students and instructors within the university, what's the point of any of this? Another student wrote last semester, when I was using points and strict rules for self-assessments: "[It's] easier to learn when the instructor grades my work because they are experts on the course material … I didn't learn anything about setting goals and assessing myself." This student saw self-assessment as yet another meaningless bureaucratic task they needed to complete to graduate.

Self-assessment isn't perfect, but it does seem to help some students feel more ownership and investment in their education, which Adrienne Rich (1977) called "claiming an education." Instead of slacking off and inflating their own grades, students in my courses have consistently reported that self-assessing makes them work harder. A student who completed the points-based self-assessments last semester wrote: "I feel more responsible to do well, and I have to meet my personal expectations now rather than a professor's, so in some ways, it's a little harder."

In this way, self-assessment may be far less radical than it appears at first glance (Morris, 2018)—it can encourage students to internalize the values and expectations of educational institutions rather than challenging them. But even if the freedom and power that self-assessment provides to students is an illusion, my classroom can at least be a space where students are practicing self-determination (Tannock, 2017) rather than training to be authoritarian subjects (Linden, 2017).

Moving away from grades creates more space for learning. A student wrote this response to my first version of self-assessment: "It makes me realize, for the first time in my academic career, that grades should be a secondary concern to actually learning something and growing from a course."

It hasn't always been easy for me to start by trusting students, but I've realized that it's something I need to work towards if I want to help them focus more on "actually learning" than on grades.

On Silence

Audrey Watters

I cracked open my copy of *The Cancer Journals* by Audrey Lorde (2020) this morning to reread "The Transformation of Silence into Language and Action."

The essay contains one of the quotations for which Audre Lorde (2020) is best known: "**Your silence will not protect you**" (p. 13). That sentence, even pulled out of context, is powerful—a reminder, a rejoinder, to speak.

But in the context of the entire essay—a beautiful essay on breast cancer, mortality, fear, race, visibility, and vulnerability—Lorde (2020) offers so much more than a highly quotable sentence on the responsibility or risk of silence or speech.

An excerpt:

> I am afraid—you can hear it in my voice—because the transformation of silence into language and action is an act of self-revelation and that always seems fraught with danger. But my daughter, when I told her of our topic and my difficulty with it, said, 'tell them about how you're never really a whole person if you remain silent, because there's always that one little piece in-

side of you that wants to be spoken out, and if you keep ignoring it, it gets madder and madder and hotter and hotter, and if you don't speak it out one day it will just up and punch you in the mouth.'

On the cause of silence, each one of us draws her own fear—fear of contempt, of censure, or some judgment, or recognition, of challenge, of annihilation. But most of all, I think, we fear the visibility without which we also cannot truly live. Within this country where racial difference creates a constant, if unspoken, distortion of vision, black women have on one hand always been highly visible, and so, on the other hand, have been rendered invisible through the depersonalization of racism. Even within the women's movement, we have had to fight and still do, for that very visibility which also renders us most vulnerable, our blackness. *For to survive in the mouth of this dragon we call america, we have had to learn this first and most vital lesson—that we were never meant to survive.* Not as human beings. And neither were most of you here today, black or not. And that visibility which makes you most vulnerable is also our greatest strength. *Because the machine will try to grind us into dust anyway, whether or not we speak. We can sit in our corners mute forever while our sisters and ourselves are wasted, while our children are distorted and destroyed, while our earth is poisoned, we can sit in our safe corners as mute as bottles, and we still will be no less afraid.* (pp. 14–15, emphasis added)

I have wondered this week—aloud, on Twitter—about silence.

I first heard about the shooting death of 18-year-old Michael Brown, a week ago, on Twitter. I heard about the death from a fellow writer, Sarah Kendzior (2014), who lives in St. Louis and has written extensively about the economic struggles of the city. I watched as the story unfolded on social media via the various social justice activists I follow (broadly speaking, I follow three groups on Twitter: educators, journalists, and social justice activists); journalists responded much more slowly, eventually picking up the story as a militarized police force tear-gassed an angry and grieving community. And then there were those who were silent.

Another dead Black man, just a week after the New York City medical examiner's office ruled that the death of Eric Garner—killed when a police officer put him in a chokehold—was a homicide.

Another and another and another (Lee, 2014).

Clearly social media has become an important tool that is reshaping how "the news" is told and shared. Although what happened to Michael Brown and what has happened since in Ferguson touch on some of the most important stories that the U.S. must face right now—poverty, racism, police violence—I'm not sure Brown's death would have received national media attention had people not been taking and sharing photos on their cellphones. Talking, tweeting, retweeting, amplifying at first the voices of the community of Ferguson and then amplifying the responses as "the whole world was watching."

From Audre Lorde (2020) again:

> Each of us is here now because in one way or another we share a commitment to language and to the power of language, and to the reclaiming of that language which has been made to work against us. In the transformation of silence into language and action, it is vitally necessary to teach by living and speaking those truths which we believe and know beyond understanding. Because in this way alone we can survive, by taking part in a process of life that is creative and continuing, that is growth.
>
> And it is never without fear; of visibility, of the harsh light of scrutiny and perhaps of judgment, of pain, of death. But we have lived through all of those already, in silence, except death. And I remind myself all the time now, that if I was to have been born mute or had maintained an oath of silence my whole life long for safety, I would still have suffered, and I would still die. It is very good for establishing perspective. (pp. 15–16)

Why are we silent? When are we silent? What does our silence mean? When is our silence about our fears, our vulnerabilities? When is our silence bound up in our privilege of not having to speak? When is our silence complicitous?

There is no single answer here, of course. I don't necessarily translate silence as "indifference." Silence is personal, and silence is complicated. But, as Lorde (2020) reminds us, our silence will not protect us.

I worry a lot about the silence on issues of race and gender among educators, particularly those in ed-tech. This isn't simply a matter of silence this week, silence on the death of Michael Brown or the death of Trayvon Martin or the death of Jordan Davis or the death of Renisha McBride. Yet the patterns of silence are there.

I want us to think: How might education technology—its development, its implementation—be shaped by these patterns? To act as though new technologies (be they Twitter, iPads, Google Classroom, or "personalized learning software") are free of ideology or are equally or necessarily "liberatory" seems so dangerous.

So yes, I often feel that I have to be even more vocal because silence is—has been—so deafening. It is—has been—the norm, a reflection of the privilege (white privilege, class privilege, male privilege) of much of the ed-tech community.

So yes, I am louder. I take risks—often fumbling with my words as I try to channel my frustrations—and I try to take responsibility for what I do or say (and don't do or say). I do stop and think about when and why my words are seen as "attacks."

"Attacks." There are dead bodies, and yet we talk about anger on Twitter as "attacks."

"Attacks." See, I worry about my own safety. And I worry about the safety of my allies. I worry about the safety of students of color. I worry about the safety of communities of color. Physical safety. Mental well-being. Our future.

I worry who our silence, what our silence, might protect.

What is our responsibility to speak? As educators? As parents? As citizens?

When must we force ourselves to take risks—"transforming silence into language and action"—knowing that we might fuck it up and say something less-than-perfectly-crafted, less-than-perfectly-wise. Knowing perhaps too that, thanks to social media, our voices are louder, and our platform is larger. Recognizing even that the risks of speaking, for those of us with privilege, are smaller.

If we don't speak, if we don't prompt one another to speak, then yes, we are left with silence. Where has that gotten us so far?

> We can learn to work and speak when we are afraid in the same way we have learned to work and speak when we are tired. For we have been socialized to respect fear more than our own needs for language and definition, and while we wait in silence for that final luxury of fearlessness, the weight of that silence will choke us.

> The fact that we are here and that I speak not these words is an attempt to break that silence and bridge some of those differences between us, for *it is not difference which immobilizes us, but silence.* And there are so many silences to be broken. (Lorde, 2020, p. 16, emphasis added)

A Soliloquy on Contingency

Joseph P. Fisher

I don't share the sheer outrage that some adjunct professors are directing at the tenured ranks. I really do believe that the majority of tenured faculty—I obviously can't speak for all of them—want every professor to be offered the benefits that were once the norm for university professors: stable employment, resources, research leave, health care, etc. I do believe this. However, I would be lying if I didn't admit that I sometimes bristle when I am forced to gape at the wide divide that separates me from those very, very few of my peers who have been fortunate enough to get on the tenure track.

To make a living wage, I have to work something between three and five jobs (the number changes slightly from year to year depending on how frequently I'm told mere days before my class starts that it has been cancelled). As a result, I cannot devote the requisite amount of time to research that would make me even remotely competitive for a tenure-line position. If I were to "sacrifice" some of my income to do that research, I wouldn't make enough money to pay my bills; moreover, given the hyper-competitive nature of the academic job market, there is no guarantee that my sacrifice would ever result in forward professional movement.

So, social media being what it is, there have been a lot of occasions where I am treated to Facebook status updates from my full-time peers that feature pictures of frothy lattes positioned next to a laptop or a tablet with captions like, "Caffeinating and researching in Geneva #sabbatical." Those moments make me feel jealous, and they make me bitter, because they serve as stark, disheartening reminders that my "career" as an academic ended with the completion of my doctorate. These days, I am not offered the opportunity to teach what I was trained to teach—American literature—and it is unlikely that I will ever again be given that opportunity, all of which makes me wonder, every single day, what I could have possibly done wrong to be so emphatically disowned by the profession that reared me. And no, simply using the "hide post" feature accomplishes nothing, because it doesn't alleviate the misery of feeling like a professional failure.

What can be done to assuage some of these tensions—to alleviate them before they result in the unhealthy infighting that we witnessed after MLA 2014 (Bérubé, 2014)? I suppose a starting point would be for those of us who are off the tenure track (but wish to be on it) not to allow our anger and jealousy to warp our criticisms of the profession to the point where they become wholly unreasonable. Certainly, anger and jealousy are justified. But those emotions are not rational, and it doesn't make any rational sense, in my opinion, to become enraged when discovering that a famous tenured professor had the good fortune of spending a weekend in a nice hotel that charges too much for granola bars (Patton, 2014). If we're fighting over granola bars, we've already lost the war.

At the same time, the tenured ranks, I think, can recognize our jealousy for what I just said it is: slightly irrational, but not entirely unjustified. In early 2014—on Twitter, on education message boards—I saw the term "tenuresplaining" gain popularity among contingent faculty. The term, as I understand it, is meant to describe the defensiveness that full-time faculty express whenever their (comparatively) secure and stable academic lifestyles are criticized for being built on the backs of part-time laborers. Not ever having been the subject of tenuresplaining, I can't speak with any specificity about this particular brand of defensiveness. However, at the very least, I can suggest fewer latte pictures, a little less reiterating how "busy" your semester has been with all of the talks that you've been invited to give, and absolutely no more furrowed brows or looks of disdain when adjunct faculty say that they don't want to "move anywhere" for a job or that they don't want to live apart from their spouses or that they prefer to watch football on Sundays rather than spend those days in the library. Just because the job market has cruelly demanded this kind of transient asceticism in the past doesn't mean that job seekers in the present should continue to stand for it.

So, you know, mutual respect would be nice.

As far as actual action might go—and I know that everything I'm about to say is going to sound ridiculous—we just need to stop playing the game. I agree wholeheartedly with a column by Lee Skallerup Bessette (2013) on institutional loyalty. By definition, contingent faculty see no loyalty from their institutions. They, in turn, should show no loyalty back. When contingent faculty are offered jobs that would force them to stop teaching mid-semester, they should stop teaching mid-semester, no questions asked.

Furthermore, when search committees do not notify job candidates of their candidacy in a timely fashion (Schuman, 2013)—a month out from the interview convention (at a minimum)—those candidates should demand the option of Skype interviews or just not interview at all. None of us should be forced to pay the escalated costs associated with last-minute travel arrangements because search committees were, of course, too "busy" to evaluate applications efficiently.

Now, as far as the professional organizations go, again, I'm guarded. Those associations have recently borne the brunt of misguided and very public vitriol from a vocal subsection of part-time professors. Whatever the failings of these associations might be, the fact remains that they are powerless to control policy at every single college and university in the entire world. That's not even the role of these associations in the first place. Like Michael Bérubé, I really don't think that arguing "the MLA didn't do enough of [X]" gets us anywhere, because the Modern Language Association (MLA) can't just swoop down on a campus and right all of the wrongs meted out by a dysfunctional job committee (or any other dysfunctionality).

However, I will say that this year, for the first time in my seven postdoctorate membership years, I paid my MLA dues based solely on the scale appropriate for my teaching salary. In the past, I have always paid my dues based on my combined income (again, I work a bunch of jobs, all of them academically oriented). This year, though, I subtracted out the income that I receive for the administrative work that I do, and I paid only according to my adjunct salary, which decreased my contribution by a discernible amount. I urge everyone to do the same thing, if you haven't already been doing so. Pay these associations exactly what they should be paid, and nothing more.

Additionally, tenured and nontenured faculty should continually coordinate their efforts to call foul ball on the notion that university funding does not exist to convert adjunct professors into full-time employees—or, if nothing else, to pay part-time professors living wages. It's disingenuous at best for universities to claim that they "don't have funding" to support their faculty.

They do have funding. Universities simply choose to use that funding in ways that, quite often, have nothing to do with professors.

Despite an apparent "lack of funding," university bureaucracies ballooned in the latter part of the twentieth century (and they continue to do so today). Interestingly, this ballooning has happened simultaneously with the shrinking of the tenured ranks and the increased reliance on contingent labor. As senior administrators have eroded the tenured ranks, they have somehow managed to find enough funding to employ armies of provosts, who usually make much more money than even the most senior faculty, constitute countless university offices devoted to "assessing" student and parental "satisfaction" "metrics" and other such corporate nonsense, and, of course, building absurdly extravagant dormitories where students (no lie) can arrange for things like maid service in their dorm rooms (Olkon, 2009). Without question, the money's there. University administrations just don't want to spend that money on education.

Therefore, I can say that I wholeheartedly support slashing administrative budgets and reallocating those funds back to academic affairs. Obviously, faculty senates should have a large say in how that kind of reallocation should happen. (On a related note, tenured faculty—at departmental and university governance levels—need to do a much better job at allowing adjunct participation in governance decisions.)

I'd also like to see administrative positions be inhabited by people who hold discipline-based doctoral degrees. I very adamantly believe that universities should be run primarily by people who have academic training, not by people whose training is in the art of growing a bureaucracy. Again, amazingly, administrators find jobs for people who get degrees in university administration. All of a sudden, positions exist that come complete with stability, photocopy machine access, health benefits, private offices, prestige, and dignity. Meanwhile, in the composition department, four professors are crammed into one office and are sharing one (frequently broken) stapler. There need to be people in positions of administrative power who take these inequalities seriously and who understand that, perhaps, student satisfaction would genuinely rise if universities saw caring for the faculty as a primary responsibility.

The grim reality, as far as I see it, is that the system is irreparably broken at every level. It cannot be fixed. We should stop trying to fix it and should let it collapse. If every adjunct professor immediately stopped teaching, the American university system would instantaneously crumble. I can't even believe I'm about to say this, because it's totally naïve and impractical, but we should let that happen. Let the current system become a thing of the past so that we might build a new one for the future, a future where we won't be forced to do so much shouting at each other—and, I can only hope, to ourselves.

N=1: A Social Scientific Inquiry into Happiness & Academic Labor

Ioana Literat

Abstract

This study aims to assess the professional perspectives of Ioana Literat (hereafter referred to as 'the subject'), a fourth-year PhD student at a major U.S. research university. The sample (N=1) was analyzed using both quantitative and qualitative methods, including statistical analysis, detailed interviews with the subject, content analysis of emails and social media activity, and dream interpretation. In spite of unresolved anxieties, the data indicates a positive trend in the subject's development, while pointing to the larger challenges of pursuing a PhD in an era of contingent academic labor.

15.1 Introduction

A young scholar's doctoral education is a quintessential period for both personal and professional development. While pursuing a PhD can be an immensely rewarding experience, it also presents frequent occasions for soul-searching to those that dare to tread down this path. In addition, recent developments in the academic labor market have exacerbated doctoral students' concerns regarding their employment prospects and, consequently, their self-worth and, ultimately, everything else in their lives. However, the impact of the PhD experience on students' self-esteem and career perspectives has, surprisingly, received too little attention in the literature so far. The present study uses the convenience sample of Ioana Literat in an attempt to fill this lacuna, and to contribute to the knowledge regarding young scholars' paths to personal and professional success.

This study aims to address three consequential research questions:
1. Does an academic career represent a good career fit for the subject?
2. Will the subject be happy in her chosen career?
3. Can the subject can make a meaningful contribution through teaching, research, and service?

While it may seem that the first and second research questions are too closely related, rightness of fit does not always correlate with personal happiness, especially in the presence of external factors (such as financial and social well-being) that might influence the latter variable. Furthermore, rightness of fit is understood for the purpose of this study as a more rational and objective determinant, while perceived happiness is a subjective variable, open to the emotional meanderings of the study participant.

The three hypotheses relating to the research questions above are:
1. A PhD in Communication is a good career fit for the subject.
2. The subject will be happy in this doctoral program.
3. The subject can make a meaningful contribution to the field of communication through dedicated teaching, service, and a socially conscious research agenda.

15.2 Methods

15.2.1 Sample

The sample ($N=1$) used in this research study is a female 27-year-old doctoral student at a major R1 university. In completing the demographic section of the questionnaire, when asked to indicate her ethnicity, the subject chose 'Other.' In terms of self-reported income, the subject left the answer to that question blank.

15.2.2 Research Design

The present study includes both qualitative and quantitative methods, in order to ensure a comprehensive assessment. While this type of triangulation aims to increase the validity of the research findings, it is also meant to capitalize on the author's unique position of having unfettered access to the subject's thoughts, fears, hopes and dreams.

The qualitative methods employed in this study consisted of in-depth interviews with the subject, content analysis of emails and social media posts, and dream interpretation. The quantitative procedures, then, consisted of administering a comprehensive questionnaire to the subject, with her responses being analyzed with the aid of an extremely expensive and non-reimbursable statistics software.

15.3 Results

According to the data extracted from the interviews and the survey, the first hypothesis ('A PhD in Communication is a good career fit for the subject') was indeed supported. An unrotated factor analysis of the survey data reveals six significant subfactors that cumulatively account for 85% of the variance. The factors that correlated positively with rightness-of-fit were: the subject's passion for teaching, her intellectual curiosity, the desire to work in an intellectually stimulating environment, and, finally, the respect she has for her chosen doctoral program and faculty. The combination of these particular elements seems to reassure the subject that she has chosen the right career path.

However, the factor analysis also revealed two elements that negatively correlated with the subject's perceived rightness-of-fit. These were: the lack of creativity inherent in high-level academic work, and, respectively, the abstract nature of scholarly research. In regards to the latter, it appears that the subject cannot help but compare her current academic activities with the nonprofit work she had previously been involved in, as the field coordinator of a digital storytelling program in central India. Nevertheless, due to the superior factorial weight of the four positively correlated items, we conclude that the first hypothesis ('A PhD in Communication is a good career fit for the subject') was supported, and we predict that evidence for this statement will only increase in the near future.

The second hypothesis stated that the subject would be happy in her chosen doctoral program. At the time the research was conducted, we found only limited statistical support for this claim; however, we were equally unable to prove the null hypothesis ('The subject will not be happy in this doctoral program'). The factor analysis identified two major elements that were positively correlated with projected happiness levels: the appeal of intellectual pursuit, and, respectively, an appreciation of personal independence. The items that correlated negatively with both current and future happiness levels were: concern about her future employment prospects in the academic job market, the presence of material concerns, and a strong fear of failure. Of these, the most pronounced factor—anxiety about future employment prospects—is exacerbated by the subject's status as an international student in the United States. Non-citizenship, indeed, poses further problems for finding an academic job; adjuncts are not offered work visas and must leave the country. Considering these circumstances, the subject realizes that landing a tenure-track job may be the only way to stay in the country that she now calls home, and this realization adds a further dose of anxiety to her thoughts about the future.

Finally, the third hypothesis, concerning the subject's potential contributions, was supported by the data in this study. Beyond her passion for teaching, a major factor in this area seems to be her desire to blend communication research with prosocial engagement, materialized in her interest in education and participatory cultures. In addition, during the most emotionally intense parts of her interviews, the subject repeatedly mentioned her commitment to 'give something back' to her home country of Romania. Nonetheless, in spite of the subject's emotional sincerity on this matter, it is this author's sense that the subject's desire is based on an elusive patriotism anchored in nostalgia, faded illusions, and stubborn memories. Interestingly enough, as a conclusion to these findings, it also appears that the subject's acute awareness of these potential academic and social contributions acts as a moderator variable, increasing the strength of the relationship between rightness of fit and happiness levels, as illustrated by the above-mentioned findings.

15.4 Discussion

The implications of these findings are multifold. Principally, they point to the internal complexities associated with pursuing a doctoral degree at a rigorous American university. While a PhD is a highly coveted and respected educational degree, it also poses vital challenges and can be emotionally consuming. These challenges are particularly pronounced for international students, whose success in such a program is often hindered by employment concerns and a lingering feeling of displacement.

However, this study presents many limitations and does not claim to be a representative portrait of doctoral students' perspectives towards their program or academic field. The sample ($N=1$) is extremely small, and, due to a variety of reasons, is not representative of the general population of PhD students. Upon meticulous review of the statistical data, we have also identified a constant moderator variable (the subject's general disposition as an optimistic but highly introspective individual), which might have skewed some of the results, particularly in regards to the second hypothesis.

In terms of avenues for future research, a significant conclusion that emerges is the need for a more longitudinal study, which measures the subjects' development over time. The data obtained here is extremely time-specific, and is thus subject to the emotional and intellectual fluctuations that characterize such a transitional period in one's professional development. Nevertheless, it is the author's hope that the present study will contribute to a better understanding of the multitude of factors that shape a PhD student's career outlook. Further experiments may be conducted, in a like fashion, should others wish to apply these research questions to their own personal contexts. In spite of the small(est) sample used in this research study, we believe that the current findings may resonate within the academic community more widely, and we invite others to join the conversation.

From Ph.D. to Poverty

Tiffany Kraft

Another Ph.D. just applied for unemployment. I haven't received any benefits because my claims are under review while the Employment Security Department determines reasonable assurance of reemployment. Per my contract with one college (I work for four institutions): "This memo is not a contract for employment and may be rescinded should the class(es) be cancelled or for any other reason." Standard non-contract language of institutions nationwide, and not oblique: There is no reasonable assurance of employment for adjuncts.

My personal low and itinerant "profession" stems from a labor crisis in higher ed that's attracted the attention of unions (Flaherty, 2014) and Congress (House Committee, 2014), but nonetheless persists and perpetuates a unique poverty that affects the majority of academic laborers. And because we look forward to new email memos from colleges offering non-contractual, temporary appointments, we lesson plan, design LMS content, and draft syllabi without pay. These working conditions are disruptive, cyclical, and intentional.

It baffles me why, in a higher-ed system that holds political but not ideological power over its workers, we don't object to our labor conditions *en masse*. There are several strong voices in the argument for adjunct labor reform (Cottom, 2014; Pryal, 2013), but the more widespread false consciousness that accepts, complies with, justifies, and administers exploited labor is shameful. It would be different if higher ed wasn't posing as something it isn't, namely: an institution founded on key phrases such as *Learning and Discovery, Access to Learning, A Climate of Mutual Respect, Openness* and *Reflection* and *Community* and *Civic Engagement*. These core values are at odds with the toxic reality.

In particular, a climate of mutual respect implies shared governance, voice, and reception for all; but this is not the case. Adjuncts won't get a seat or an iota of mutual respect at the bargaining table without union representation. Of course, murmurs of unionization are met with resistance and censure, and there isn't an internal path of negotiation for intentionally marginalized university workers.

Imagine a university where unified faculty teach and write with the dignity and pay our work deserves, administration prioritizes instruction and essential student-support services, and student-centered learning models are progressive, not packaged. Until then, it's time to stop romanticizing a bygone academy and, rather, court new paradigms that are proven, ethical, and sustainable (Vancouver Community College Faculty Association, 2013). Higher ed needs radical leaders who realize their role and stake in the crisis, quash cronyism, and confront the culture of fear and contempt that hamstring progress.

It's not unusual for adjuncts to spiral in a climate designed to exploit and scapegoat, which Colman McCarthy (2014) reiterates: "the demeaning of adjuncts is little more than structural economic violence." Unemployment isn't a choice, it is a national security "designed to provide partial income replacement to regularly employed members of the labor force who become involuntarily unemployed" (Office of Research, Evaluation, and Statistics, 1997). For me, going on the dole is a last resort, a demeaning consequence.

Even temporary poverty is difficult to bear; it's humiliating and gut wrenching. That said, this personal crisis has humbled and brined me in reality, and I am determined to fight for my profession, from the margins or beyond, if with luck and effort I should get a job outside academe through weekly job searches. It isn't that simple, though. I've been in a vacuum for ten years, teaching toward tenure (yeah, I know), and numb to change because I was employed, however insecurely. I wish I'd bartended in my twenties so I could delete the Ph.D. off my CV, take away the M.A., bury the B.A. under bar-back experience, and get a job with tips pulling pints and shaking martinis. It sounds nice, doesn't it? But I'm not sure I can wipe the slate or deny experience at my age.

In talking to colleagues in similar crises, it's apparent that the slow erosion of the profession has taken a toll, and though there will not be a mass exodus of adjuncts, there are hordes of us who, battered by academe's hard-labor mills, contemplate alt-ac careers. And those who break out often reflect on their precarious employment with the fermented retrospect one affords a broken marriage.

I'm in this argument to humanize contingency and reify the argument for adjunct labor reform. The narrative's shifted: it's time to *realize* change and fortify higher ed for future generations. Incremental strides are within reach when we work collectively toward a solution, with mutual gravitas. It's time to pick up a spade and cultivate a voice and conscience, like in this exchange between me and Nathaniel C. Oliver (2014):

> Thank you, Nathaniel: Thank you for bearing all. I've adjuncted for a decade, and the pain compounds with every narrative I read. This is my story, your friend's story, and so many others.
>
> The image of a sardine-packed train with pushers pushing more bodies in the closing doors comes to mind. Will it derail? Will I get on? Should I bother? I have a ticket, though!
>
> I don't want to die an adjunct, either, and I realize this pathway to poverty is secure. I have a little fight left in me, though. And each act of bravery, such as your post, ignites my passion and purpose to push harder. I'll get on that fucking train and ride it. Thanks for the push.
>
> …
>
> Thank you, Tiffany. I am continually impressed by the determination that people like you have shown in the face of the juggernaut that is adjunctification, and I am just trying to do my part. I understand that the temptation is huge to just give up completely, especially now that I have a young daughter, and like all parents, I want to provide for her the best way that I can. On the other hand, I worry about the world that she will inherit; I don't want it to be one where academics are routinely consigned to poverty while others make fortunes off of their labor.
>
> Good luck to you, and thanks for not giving up the fight.

For Nathaniel, for my children, for my colleagues, I won't give up. I have so little to lose but integrity and you. Remember when we were kids on recess? Some chased, some swang, some played four-square? It's quite the same now: The playground's changed, but we're on the same merry-go-round. Now, as then, I'm against bullying. The bell's calling us out…

When One Class is Not Enough

Amanda Licastro

It is only from a place of extreme privilege that someone is able to shelter themselves from politics, as many must face the consequences of the political system in everything that they do. This is especially true of educational institutions, in and through which structural inequality is systemic, unfairly impacting students, faculty, and staff of color. In my 2017 article for *Hybrid Pedagogy*, "Learning at the Intersections", I outlined an approach to teaching highly politicized topics through service learning, but more specifically I explained how I taught a course on the issues of citizenship and what I would do differently in the future. Now, in 2020, the week after one of the most contentious elections in recent history, I find myself reflecting on the evolution of this project, and how my approach to bringing politics into the classroom has changed. Our country is deeply divided, and as a result of—or perhaps at the root of—that chasm there are certain "hot button" topics that are sure to ignite controversy. It may seem easier to avoid these topics, or navigate course conversations away from these debates. However, in reimagining my courses, I have tried to meet the challenge of discussing political topics transparently and aggressively to frame a space for students to engage in the much needed discourse of social justice through the lens of rhetorical empathy.

As Lisa Blankenship (2019) writes in *Changing the Subject: A Theory of Rhetorical Empathy*, "from its beginning, empathy has signified an immersion in an Other's experience through verbal and visual artistic expression. This element of an immersive experience that results in an emotional response, as well as the associations of empathy with altruism and social justice, possibly explains its continued linguistic cachet over terms such as pity and sympathy" (p. 5). In my courses we study the stories of American immigrants, refugees, and asylum seekers, and I try to move the discussion away from the feelings of sympathy at best, or indifference at worst, to a place of altruism. I say "courses," because at the time of my first article this work was isolated to my Digital Publishing (200-level) class, but now has become central to my Introduction to Literature (100-level) and Grant Writing (300-level) classes as well. Teaching this content at three different levels to a variety of audiences has helped me hone what is essential to the success of the course, and that starts with language. I find many students are reticent to participate in conversations about controversial topics because they do not know the appropriate language to use, and they do not want to mark themselves as in opposition to anyone else in the class. In fact, in their reflections essays students lament that they wish they would have spoken up more and express similar sentiments of struggling to find their voice. Based on this feedback and my research on declining empathy rates (American Psychological Association, 2019), I started experimenting with both analog and digital tools to immerse students in the rhetoric surrounding citizenship in the United States.

17.1 Start with the Personal

After too many years avoiding revealing any personal information to my students, I realized that in order to establish my position on the topic of immigration, I needed to share my story. I start from a place of vulnerability, in hope that this openness will inspire bravery. In the first week of class, I search the free Ellis Island Passenger Search for the name of my grandfather, Gabriele Licastro, and go through the process of reading clues to fill in the details of my family history.

By asking questions of the passenger manifest provided on the site—What kind of ship were they on? Why is everyone on board male? Why is everyone on board marked as a "seaman"? Did you notice how many passengers were hospitalized?—we are able to draw some solid conclusions: My grandfather was a prisoner of war, brought over on a military transport in

1941 with a group of captured members of the Italian military. I then share that my grandfather was put in a labor camp, where he survived by cooking for the American officers, and when eventually released, found employment as a chef in Cleveland, Ohio. Students experience the act of historical research, but through the lens of empathizing with a real human, their professor, who has inherited this generational trauma.

I don't ask students to reciprocate with their immigration stories; each student determines for themself whether to disclose personal information. Instead, I follow with a low-stakes creative-writing project inspired by NPR's "Where I'm From" series (Noenickx, 2019). The results are simply gorgeous. Students write short poems about their families, home towns, traditions, fears, regrets, pain, and joy. I also encourage each student to record themselves reading their work to share alongside their poem. Especially when we were forced to be fully remote in March 2020, these glimpses of vulnerability brought a sense of community and caring to the sudden distance. Perhaps even more so than in the face-to-face classroom, inviting the personal into a virtual learning space breaks down the walls of academic discourse and helps students find their voice.

17.2 Interrogate the Public

After looking inward, my classes turn to representations of citizenship in popular media sources. I ask students to find sources from the last six months that focus on issues faced by immigrants, refugees, and asylum seekers. I encourage students to find articles from sources they are likely to read outside of class. The goal here is to capture a wide variety of sources to analyze and compare together. To my surprise, students rarely duplicate articles, which suggests the class is diverse enough to retrieve different search results, or have different filter bubbles, which are issues that we discuss in class together.

Then, in collaboration with Sara Godbee, the subject librarian embedded in my courses, we guide the students on a lesson in evaluating bias and identifying loaded rhetoric in news publications. This activity could be easily adapted for any topical issue, and can be done remotely or in person. Students locate the publication they selected on the widely-seen "Media Bias Chart" (Ad Fontes Media, 2021), and then label their article as left-, center-, or right-leaning.

Once the articles are labeled, we create a "gallery walk" (shamelessly borrowed from Dominique Zino, Associate Professor of English at LaGuardia Community College), by posting the articles around the room—which can be done virtually via a discussion board or blog. Students identify and plot words that convey urgency or bias from the left/center/right, using polling software online (which could be done with sticky notes in person) to track word frequency. We use the visualizations to draw conclusions about language used in journalistic reporting. We discuss how terms like "crisis" appear across both sides of the political spectrum, but labels such as "alien" and "victim" are polarized. Inevitably, there are surprising results that contradict our expectations, allowing both the students and myself to re-evaluate our assumptions. Thinking through why we expected certain terms to be associated with the left or right helps us uncover our own political bias.

The process is repeated with articles from academic journals, with the timeframe expanded to the last five years to accommodate scholarly publishing timelines, which demonstrates how language changes over time and in different disciplinary contexts. In a short reflection assignment, students consider how language is welded politically—even in scholarly publishing—and how this exercise prepared them to enter a specific discourse community. The exercise prepares them to distinguish between—and enter into—both the academic and service communities. I put pressure on students to consider why it is important to go beyond the concept of what is "politically correct" and instead investigate what terms they feel comfortable using as an individual now that they have examined the rhetorical strategies of multiple sources.

17.3 Finding Truth in Fiction

Perhaps no one better contextualizes the problem with language than Chimamanda Ngozi Adichie. In my 2017 article, I concluded that the course would be better served by integrating fiction to provide students with more diverse narratives to connect to, in particular, Adichie's 2014 novel, *Americanah*. That has proven undeniably true. I've introduced this novel into two separate courses, alongside her 2009 TED Talk on "The Danger of a Single Story," pairing it with a rotating selection of scholarship, poetry, and short stories that fit the needs of that year and student population. For example, in my Fall 2020 Introduction to Literature course, which serves as the only Intercultural Knowledge Competency course in the university, I've used a combination of the following resources:

- I show the videos for "Immigrants" from The Hamilton Mixtapes performed by K'naan (Hamilton, 2017) and "This is America" by Childish Gambino (Glover, 2018) paired with the annotated lyrics via genius.com. These sources serve as an introduction to close reading visual and textual representations of immigrant narratives.

- We collaboratively annotate poetry by Langston Hughes, Jimmy Santiago Baca, and Hai-Dang Phan using social annotation software, specifically hypothes.is.
- We consider short stories by Carmen Maria Macado and the graphic novels of Kate Evans, and I ask students to analyze small sections of the text they find provoking.
- I bring in Virtual Reality (VR) immersions such as "Limbo" (The Guardian, 2017), "The Displaced" (Solomon & Ismail, 2017), and "Clouds Over Sidra" (Arora & Milk, 2015), as well as critiques of the medium from Janet Murray, Liz Losh, and Lisa Nakamura.

In each text, we find moments of familiarity, points of connection that help us identify with the characters, but then also points of departure that highlight our differences and make the texts challenging. As we traverse these tensions, the complexity of empathy is apparent; we have a hard time understanding those we see as Other. While all of these texts are difficult for students to navigate, the conversations are rich and full of the kind of exploration needed to reframe representation as historical, cultural, and institutional.

The results are some of the most poignant student work I've witnessed in my ten-year career as an educator. Nothing seems to resonate as strongly with students as *Americanah* (Adichie, 2013). Therefore, I've made her novel the anchor text for my "remix" midterm project in which students select one character and tell their story in a new medium. In groups, students create podcasts, video games, comics, or VR experiences that require them to research the political and social issues at work in the narrative thread they chose. By collaborating together, and choosing roles that best fit their strengths, students break down the work into manageable chunks and help refine the overall concept to an achievable vision. Peer review sessions and conferences with me also aid in avoiding pitfalls and troubleshooting technical difficulties.

One group, compelled by a slew of articles about black students being discriminated against at school because of their hair styles, made a podcast episode titled "Neither Hair Nor There" (Kramer et al., 2019) in which they interviewed two of the female characters about their experiences trying to find products to take care of their natural Nigerian hair in America, and interspersed references to the articles by way of asking the characters to reflect on the prejudice they faced. They ended the podcast post with real resources to help black women find legal language to advocate for themselves as well as natural products to maintain a variety of hair textures.

Another group created a stealth-mode VR simulation (Fedor et al., 2019) set in London which traced Obinze's plight of trying to remain unseen as an illegal immigrant after his student visa expired. The player has to wear a variety of masks as he assumes different identities through borrowed or stolen documentation. The goal is to remain unseen while still working, traveling on public transportation, buying food, and entering into an arranged marriage for the purposes of gaining citizenship. This method allows the player to witness the unending stress and anxiety faced by millions of undocumented workers around the world.

And yet another group used Twitch to design a decision-based game (Perry et al., 2019) in which the player decides whether or not to engage in illegal activities while trying to survive as a legal Nigerian immigrant in the United States. The sheer number of branches their decision tree had—and the images they chose to embed throughout—impressed me, but the act of forcing the player to face these decisions was the most powerful part of this game. The game offers players a view into the authentic tragic and unjust trade offs our society sets before immigrants.

Blankenship (2019) defines empathy as "both a conscious, deliberate attempt to understand the Other and the emotions that can result from such attempts—often subconscious, though culturally influenced" (p. 7). As I hope you can see, these projects require a great deal of academic rigor while making space for emotional and personal connection to the material. Based on Collaborativism from Linda Harasim (2017), this constructivist approach to pedagogy emphasizes peer-to-peer learning, allowing those who may feel marginalized or underrepresented in a classroom to be in a position to lead or share expertise. Classmates share their diverse experiences and perspectives, stimulating difficult conversations but also innovative problem solving. Of course, some groups stumble, projects stall, and individuals disagree, but those efforts are valued even if the final product is not perfect. The point is to experiment with the unknown and apply what they have learned to make something new. The goal is to better understand the experience of asylum seekers through the lens of rhetorical empathy.

17.4 Serving the Community

Rhetorical empathy is particularly important in my service-learning courses, where students work directly with asylum seekers. Understanding the language students use to address and describe our clients ameliorates the apprehension in these encounters. Students enter the immersion with the tools needed to be thoughtful and empathetic in their interactions with the staff, volunteers, and asylum seekers themselves. As I learned from my previous experiences engaging in service-learning work, it is important to have multiple community partners in case situations arise that make the collaboration impossible to complete. Therefore, I now invite three or four organizations to present to the class, and students choose which they feel passionate about serving. This allows students to match their interests to the coursework, and gives each group a clear motivation for the final project.

The final project varies based on the course objectives. The 300-level students draft real grants to funding agencies, the 200-level students create digital products for the organizations to distribute, and the 100-level courses submit small internal proposals for one-time events or projects. Consistently students articulate the value of having real clients for their final project, expressing that the high stakes of providing resources for those in need motivated them to excel. They are also able to explain that this final project enables them to apply what they had learned throughout the semester to a tangible outcome outside the limitations of the classroom, giving them an opportunity to synthesize a variety of materials and make use of the content for the greater good. These connections are exactly what I hope to achieve as an educator. Working from Adiche's prompting, students share examples of their lived experience and develop trust and authority as they cross cultural boundaries, empathizing with teammates as they work to build empathy for the client.

In critiques of empathy, many suggest that because empathy is an emotion, it is fleeting (see, for example, *Against Empathy* from Paul Bloom, 2017). By connecting the immersive narratives students experience to the application of those lessons in the service they provide to our community partners, the empathic response moves from emotion to action—empathy, then, is a catalyst. It is precisely this movement that makes the argument from Blankenship (2019) so compelling to me and negates the common critiques of empathy.

17.5 Conclusion

As dedicated professors, after we work on a course so diligently, and thoughtfully, sometimes for years, we hope the effort has an impact. We read through the glowing reviews of our students and accept the praise of our colleagues with satisfaction. However, there are also times when we must face the unexpected results, the comments that, despite our very best intentions, prove that there is more work to be done. In spring 2020, my 100-level Intercultural Knowledge Competency course received the kind of student evaluations we all long to read—except one. One that stated, "I wish it broadened the demographics of immigrants besides African Americans." That anonymous comment gutted me. After three years of carefully crafting a beautifully diverse set of voices to uplift my students and make space for sharing and connection, I was met with the blatant realization that one course, one teacher, one program, would never be enough. I cannot cover all of the concerns, represent all of the voices, and influence everyone's perceptions in one semester.

This moment of personal failure was later compounded by the hopelessness felt nationally by educators whose intellectual, emotional, and physical labor was erased by a political policy. The oxymoronic Executive Order on Combating Race and Sex Stereotyping resulted in the closure and cancellation of diversity programs at universities across the country (Flaherty, 2020). How can we cultivate empathy in our classrooms when the national dialogue contradicts our efforts through a much louder, much more powerful platform?

My answer for you is the same philosophy I use in my classrooms: collaboration and community. We must find strength in numbers, because one classroom, one teacher, one program will never be enough. But if we commit to making every syllabus, every curriculum, and every college more diverse, our platform will be amplified. And further, if we connect those learning objectives to service learning, and engage in outreach with our students, the message will be too loud to drown out.

Assessing so That People Stop Killing Each Other

Asao B. Inoue

> "Is it possible to teach English so that people stop killing each other?" Ihab Hassan asked my group of teaching assistants in 1968. We are still trying to come up with an answer.
> —(O'Reilley, 1989, p. 143)

I open this chapter[1] with Mary Rose O'Reilley's invocation of Ihab Hassan's question for several reasons. The short 1989 article in which O'Reilley offers the above is a kind of rumination on her teaching life to that point, which began in the 1960s. She asks, "how did I get here," and invokes the infamous article and book by Jerry Farber (1968), *The Student as Nigger*. The question above is prompted by a growing cynicism in her own teaching, and a sense

---

[1] Originally published as "Assessing English So That People Stop Killing Each Other" in Inoue, Asao B. (2019a). *Labor-Based Grading Contracts: Building Equity and Inclusion in the Compassionate Writing Classroom*. Perspectives on Writing. The WAC Clearinghouse; University Press of Colorado.

that "young people in the profession know rather little about the history of what, to some of us in mid-career, is still 'the new pedagogy'" (O'Reilley, 1989, p. 143). The new pedagogy she speaks of is loosely the student-centered classroom and discussions of power relations in the classroom, pedagogies that look to give up power, pedagogies that agree with many of labor-based grading contracts' basic assumptions.

Labor-based grading contracts can offer students in writing classrooms the chance not just to redirect the way power moves in the classroom, but to critique power, and that begins by making obvious how power usually moves and who controls it. Labor-based grading contracts show us that in writing classrooms, power can move, not through standards and teacher's judgments of student writing—although teachers still judge writing—but through students' own labors. While they de-emphasize the dominant, White, academic, discursive standard, they may make learning such a standard easier for many students if writing with less anxiety and the ability to take more risks in writing is linked to such learning. But mostly, I promote labor-based grading contracts because they can encourage assessment ecologies that value multiple Discourses and allow students to maintain their right to their own Discourses in the English writing classroom. I promote them because they make learning a dominant White racial Discourse problematic (in the Freirean sense), offering conditions in the classroom that allow diverse *habitus* and judgments to sit side by side in tension, allowing students to question and critique that dominant Discourse while paradoxically having the choice to learn it or something else. I promote them because they work against White language supremacy by offering conditions for counter hegemonic discussions about language and judgment, and allow for alternative ways of languaging that provide students with flexible, rhetorical practices that can help them in their futures. Ultimately, I promote them because they create sustainable and livable conditions for locally diverse students and teachers to do antiracist, anti-White-supremacist, and other social justice language work, conditions that are much harder to have when writing is graded on so-called quality or by some single standard, and when students' labors are not fully recognized and valued. These conditions, conditions that I believe are fairer for raciolinguistically diverse students, open the writing classroom to ask similar questions that Hassan and O'Reilley (1989) do. And they start with standards controlled by teachers.

Do standards in English writing classrooms kill people? Hmm. Maybe a better question is this: In a world of police brutality against Black and Brown people in the U.S., of border walls and regressive and harmful immigration policies, of increasing violence against Muslims, of women losing their rights to the control their own bodies, of overt White supremacy, of mass shootings in schools, of blatant refusals to be compassionate to the hundreds of thousands of refugees around the world, where do we really think this violence, discord, and killing starts? What is the nature of the ecologies in which some people find it necessary to oppress or kill others who are different from them, who think or speak or worship differently than them? All of these decisions are made by judging others by our own standards, and inevitably finding others wanting, deficient. People who judge in these ways lack practices of problematizing their own existential situations. They lack an ability to sit uneasily with paradox.

I don't mean to suggest that there are not some cases where a person is simply mentally ill or an anomaly, the exceptions to the norm. I'm saying there are far fewer of those cases than we may think. If literacies are bound up not just with communication but with our identities and the social formations that people find affinity with, if literacy is bound up with how we understand and make our worlds, then a world with literacy classrooms that use singular standards to determine progress and grades of locally diverse students, a world that holds every student in the classroom to the same standard regardless of who they are or where they came from or what they hope for in their lives, is a world that tacitly provides and validates the logics of White supremacy. It is a world that promotes White language supremacy. It is a world that validates the use of a dominant *habitus* to make similar kinds of judgments of people elsewhere outside of school.

Our students learn how to judge their world by the practices of judgment they experience as they move through their worlds. Experiencing standards over and over in classrooms validates by repetition the practice. If standards are always applied and people are ranked based on them, if people are denied things because of them in dispassionate ways through the first twelve or sixteen years of one's life—the crucial literacy learning years—then I think it is easier to justify judging everyone, no matter the subject or decision, circumstance or situation, by a single standard, unproblematically, and those judgments lead, if one pushes the logic far enough, to killing.

So, how do we teach so people stop killing each other? Perhaps, we might ask, how do we judge language so that people stop killing each other? That, I think, is the real question. This is the exact problem that I argue labor-based grading contracts explicitly addresses in writing classrooms, the problem of grading locally diverse students, the paradox of teachers who are by necessity steeped in a White racial *habitus* while many of their students are not, the problem of how to help students and teacher confront and discuss bravely the racialized politics of language and its judgment. Yes, if we can confront such paradoxes in the judgment of language, in the judgments of *habitus* through our *habitus*, then maybe some of the killing may stop.

O'Reilley (1989) concludes her article: "The point is, you can't just put your chairs in a circle and forget about the human condition" (p. 146). I wish I could say that this good conclusion was on my mind over most of the last fifteen years as I developed my version of contracts, but it wasn't. It has only been in the last five or six years that I've understood how important it is to account for the human condition, that is, the material conditions, the embodied conditions of learning in various, diverse bodies who inhabit different places in our larger community. This human condition is implicated in any writing classroom where a group of locally diverse (or homogenous) students come together to read, write, and engage. And what is more critical to the human condition, as Hannah Arendt (2013) reminds us, than labor, work, action. No matter how one wishes to define these terms, they reference people toiling, exerting, struggling, trying, suffering, succeeding, and failing. They reference making and historicizing, building for others, not just for ourselves. Laboring, which may be a good synonym for suffering in the writing classroom, is quintessentially the human condition.

Ten years after O'Reilley wrote the above article, she revisited her teaching in *Radical Presence: Teaching as Contemplative Practice* (1998). In its opening chapter, she says, "I would like to ask what spaces we can create in the classroom that will allow students freedom to nourish an inner life" (p. 3). What she means by "an inner life" are contemplative practices that might offer students learning and something else, something human, perhaps something that acknowledges their unique human conditions. What she offers in the book are beautiful ruminations and contemplative practices from her classroom, deep listening, paying attention, being still enough to notice, standing in radical presence. Here's how O'Reilley (1998) describes the practices of deep listening from her classroom:

> It deals with the whole rather than with the parts: It attends not to the momentary faltering but to the long path of the soul, not to the stammer, but to the poem being born. It completes the clumsy gesture in an arc of grace. One can, I think, *listen someone into existence*, encourage a stronger self to emerge or a new talent to flourish. (p. 21, emphasis added)

What strikes me about O'Reilley's contemplative pedagogy is its compassion and its potential for growing the patience in teachers that is needed when we confront students who are different from us, who do not look, or sound, or come from the same places as we do, or want the same kinds of things for themselves as we do. Her pedagogy is one that asks us to listen deeply to our students, cultivating enough grace to allow for their seemingly clumsy gestures, their momentary faltering in words, so that their poems, or papers, or new selves, can be born. Labor-based grading contracts offer conditions, for such compassionate pedagogies to work, pedagogies that can, I think, listen many students into existence. Or rather, labor-based ecologies, ones fundamentally focused on the three dimensions of laboring, ones that do not use a dominant White standard of language to rank students, provide an encouraging and compassionate place for us to *attend* to our students, for students to *attend* to each other and themselves. Attending is more than an auditory metaphor. It is more fully embodied and compassionate. It includes a vital part of what I hear O'Reilley asking us to consider in our pedagogies: the material conditions of learning, living, languaging, and laboring. Attending includes the bodily, which is also about presence—being present for ourselves and others. It is about paying attention to this still moment, acknowledging the emotional and intellectual dimensions of it, and about beholding that which is becoming in front of us all the time. I believe, labor-based grading contracts help cultivate assessment ecologies in which students have more ability and more opportunity to be radically present, to be here in this moment, the only moment any of us have, and just practice.

In 1997, Fred McFeely Rogers, the acclaimed host and originator of "Mr. Rogers' Neighborhood,"[2] the public television show for children, received a lifetime achievement award at that year's Daytime Emmys. In his now-famous and short acceptance speech (The Emmy Amards, 2008), he asked the audience for a favor: "All of us have special ones who have loved us into being. Would you just take, along with me, ten seconds to think of the people who have helped you become who you are. Ten seconds of silence." I cannot think of a more compassionate way to articulate the way each of us becomes who we are today and who we will be tomorrow. But to see it, to see the loving into being, requires what Mr. Rogers asks of us, what O'Reilley asks, that we attend others into being, that attending is an act of love as much as it is of grace, and loving helps others become. As I have reflected previously (Inoue, 2019b), we are all *becoming*, in all the ways that that word can mean. I was loved into being because I was becoming. I was a beautiful brown boy, a becoming brown boy in a dark world of White supremacy and racism with just enough people around me to attend me into being, and it is my obligation to return that attending and loving, first to those who loved me into being, then to others who are not me, my students and colleagues, all of whom are becoming themselves. Is there anything more important? Is there a better answer to Hassan's and O'Reilley's question?

---

[2]*Mr. Rogers' Neighborhood* first aired in 1968 and recorded its final shows in 2000. By the time Mr. Rogers had finished, he had been awarded four Emmys and forty honorary degrees and had recorded 896 episodes of his TV show ("Fred Rogers", 2020).

While I realize that some of our students, perhaps even some of us teachers, may not characterize our childhoods as places in which people around us love us into being, but maybe we might imagine a classroom in which this could be true. A present and future that is becoming. We might think of the assessment ecologies we cultivated with our students as places that invite us, urge us, move us to love our students and their writing labors into being, to attend to them without ranking.

To attend to others into existence, to act in compassionate ways, and to be radically present are the same practices. They are labors of loving and learning, of living and growing. To love is to attend, to deeply listen to another who is not like us, to be present for them, and to do so on their terms, not to change them into our image of middle-class Whiteness, or some other *habitus*, but to simply do so because they, like us, are becoming. Love-attending is a practice of radical presence. It is not easy. But our students are here. We are here. It is now. We have no other moment but now. Really attending deeply means sitting with another in their relative suffering, being compassionate, without conditions, like our mothers and grandmothers, fathers and grandfathers, our brothers and sisters often do, or did, or could have done in a more perfect world.

In a recent First-Year Writing (FYW) course, the second in our stretch sequence, one of my students offered a description of his past literacy experiences, hinting at what our class' labor-based grading contract gave me. He is African American, with parents from Africa, but he was raised in the U.S. I leave his "stammers" and "clumsy gestures" to urge you to attend deeply right now, right here.

> My experience in the past with literacy hasn't been positive; when I was as young as I can remember when it came to writing or reading I just wouldn't do it, I didn't like it. Like in elementary school, reading especially was always rewarded. During those schooling days logging our reading for the school week was a requirement; however, if we read long enough or read a challenging book we'd earn points and could trade them in for candies, toys and electronics. But I soon compared myself to other people because of the expensive things they got from their points, which in turn I saw as them being extremely proficient at reading. So what I did was take a bunch of challenging books that were above my level and stressed myself meaninglessly over them and putting myself down because no matter what I tried, I couldn't read at the level of my peers. It all just became some silly game to me. My younger self was thinking "I only play

games that I like so I'm just not gonna go a deal with that," and for the longest time that's what I've seen it as, something that I just don't want to partake in. So I gave up. Gave up on trying to be like everyone else, and until recently only ever saw reading as a chore. This goes the same for writing too. Whenever I had to do it, it was just boring. Was always told to close read the literature, look for devices and methods in the writing. You don't know how many times from a teacher I've heard "look for the literary devices the author uses to convey their purpose." Sure it was a of learning about literature, but I thought it was a superficial way of learning; could never apply what was taught towards my own endeavors because I felt what was taught was so shallow.

Now, it isn't so bad thanks to this class when I started it in the winter quarter, it got me used to reading and writing, especially writing.

When I sit in the presence of my student's words, when I try to listen deeply, when I stop placing any of my expectations on him for this writing, I don't have to ask or urge him to find more meaning than the final sentence, than the simple fact that our labor-based grading contract ecology "got me used to reading and writing." That is something, given his past experiences. He is becoming right in front of me, and I'm lucky enough to witness it.

But this doesn't mean I cannot dialogue with him, ask more questions, and do so in an environment that rewards this extra labor. I can model a way to compassionately attend him into being, and he might return that attending to me or his peers. But he will surely see an alternative to the standards-driven, White language supremacist classroom that I'm arguing does so much harm in and out of school. He will get chances to problematize the judgments of language and consider the ways our *habitus* function in systems of judgment like those in schools, like White supremacist ones in the larger society. Such an ecology, such a writing classroom, assesses writing so that people might stop killing each other by seeing difference not as a threat or as wrong but as another becoming. Yes, I have flimsy evidence for such a claim, but if I'm going to have faith in anything that will stop the killing, and violence, and discord in the world, I'd like it to be our loving and compassionate attending to each other.

Politics

19 Pedagogy as Protest — 117
JESSICA ZELLER

20 Critical Citizenship for Critical Times — 121
MAHA BALI / مها بالي

21 Interrogating the Digital Divide — 125
LEE SKALLERUP BESSETTE

22 Pedagogy and the Logic of Platforms — 129
CHRIS GILLIARD

23 (dis)Owning Tech — 133
TIMOTHY R. AMIDON

24 Education as Bulwark of Uselessness — 143
LUCA MORINI

25 The Political Power of Play — 151
ADELINE KOH

26 Ghost Towns of the Public Good — 159
PAT LOCKLEY

27 Ed-Tech in a Time of Trump — 163
AUDREY WATTERS

Pedagogy as Protest: Reimagining the Center

Jessica Zeller

I write this in June of 2020, in the wake of the unjust murders of George Floyd, Breonna Taylor, and Ahmaud Arbery; Atatiana Jefferson and Fred Rouse where I live in Fort Worth; and just two days ago father of four Rayshard Brooks; among too many other innocents who should have been alive today.

I write this as we are still inexplicably engaged in a conversation about the humanity of Black people, as though it were somehow up for debate.

I write this as global uprisings against police violence and systemic racism are entering their third week while losing the attention of the 24-hour news cycle and those who hashtagged their way to a suspiciously visible allyship.

I write this during Pride month, as LGBTQ people's legal rights are being deliberately rescinded by a bigoted president and their identities publicly invalidated by a transphobic children's book author. As Black trans women are still being murdered and forgotten at an alarming rate.

I write this as the Coronavirus pandemic continues to escalate, disproportionately affecting underrepresented and underserved communities. Over 110,000 Americans have died, and the government has turned away—willfully negligent and criminally inept. Our national mourning has been negated by a political horror show.[1]

I write this as grief has become pervasive and accepted. As "just checking in" and "wanted to see how you're doing" have become essential daily communications with loved ones. As "I hope this finds you safe and well during this difficult time," has become the standard prologue to our emails.

I write this as I am hesitant to acknowledge my anger. My white, female, cishet identity keeps me from the prejudices, the racism, the centuries of hate. I can only try to imagine the degree of rage and the kind of exhaustion that one might feel in the face of it daily. As a woman I can sometimes relate. Sometimes. To some extent. I donate and read and call and write until the fury gives way to a less volatile feeling of existential malaise. There is too much suffering in too many places happening all at the same time; seeking out moments of joy takes dedicated effort. Doing the work helps. No one needs another white woman's tears.

Scholar and activist bell hooks (2003) is renowned for her work to dismantle what she calls "imperialist white supremacist capitalist patriarchy." These interrelated centers of power, she teaches, are responsible for the oppression and domination that shape our world. In 2010, hooks came to my alma mater as a Visiting Distinguished Professor of Women's Studies. Despite being hosted by five different departments and centers across the massive Ohio State University campus—interdisciplinary voice that she is—hooks's one public lecture took place in a not-nearly-big-enough lecture hall. We broke fire code, cramming as many people as we could into the rows and clogging the aisles. We were doubled up in seats and smashed against walls. The administrators present stood aghast, powerless. And as only a true critical pedagogue would, hooks invited hordes of students from the audience to fill the stage with her. She brought us physically together in community around her, an act of generosity which served as the precursor to an intellectual ass-kicking that brought us—if possible—even closer.

If you've ever been fortunate enough to hear hooks speak, you know it's a near-spiritual experience. She's a philosopher and a storyteller; sharing her

---

[1] At the time of revision in November of 2020, the U.S. is approaching a quarter of a million deaths from COVID-19. Projections suggest that a half-million deaths are possible by March 2021.

own narrative in a way both theoretically significant and personally meaningful. Her work to expose systems of oppression is at once about her and about all of us, collectively and as individuals in the world. It seems to reach out from multiple centers, and all at the same time. Through publicly accessible academic discourse rooted in a love ethic, she finds us where we are and shepherds us into a critical, contextual awareness of ourselves and others. Her work always feels urgent, essential, human.

In the face of so many unknowns as we approach the Fall 2020 semester, colleges and universities are treating educators like vehicles for "content delivery." They're pushing too many one-size-fits-all course models that, like department store winter gloves, don't actually fit all. There are too many cookie-cutter solutions. Too many catchphrases. Too many online platforms and learning management systems with too many biases that disadvantage too many students. And too many damn hyphenates; "standards-based" and "data-driven" among the worst offenders.

"Student-centered" might be the most misused of all the hyphenates the education field has ever devised; it's lipstick on a pig, so to speak. Ideally, we wouldn't have to say "student-centered" at all. It would be apparent in our work. It would manifest. And yet for some reason we've come to need a term like "student-centered" to remind ourselves and our institutions that there are indeed students present. In spite of our slick buzzwords and "flipped classrooms," the students are nowhere near the center. Many have left the room unnoticed.

Often occupying the center of the learning space in their stead, "imperialist white supremacist capitalist patriarchy" has taught us that students are the enemy. Our syllabi are a bloated ten pages long and thick with policy statements, as too many in education have come to believe that good teaching and rigid rule enforcement are one and the same: no late work accepted; grade deductions for late arrivals; required use of surveillance software; "fairness" as represented by uniform punishments regardless of personal circumstance or hardship.

Where is our humanity?

It's no wonder that many students seem only mildly interested in school, if at all. School isn't made for them. Not when there are accrediting agencies and state standards and educational technology contracts in play. Not when institutions rely on unethically sourced student data analyzed with questionable integrity to confirm their use of often inequitable "best practices." Not when the Ivory Tower doesn't. even. try. to listen or respond to those learning and trying to learn.

In this fraught moment, I am looking to pedagogy. I am embracing the curriculum and the classroom and yes, even the Euro-centric ballet studio as sites for resistance; places where trusting students, expressing interest in them, and giving them the benefit of the doubt are seen as radical acts that defy the "imperialist white supremacist capitalist patriarchy" that we've allowed to remain in the center for too long.

The pedagogy I see is equitable. It invites students into the center by valuing their differences, recognizing their experiences, and affirming their identities. It practices acceptance. It centers our collective humanity, asking out loud and as part of the process: Who is learning for? Who is learning about? Who authors learning? And why?

This pedagogy is responsive. It prioritizes checking in on loved ones, holding space for grief, and honoring rage at injustice. It situates learning in and through and with community. It notices how we communicate; it listens when we speak. It is at once about our individual stories and about us, together, in the world.

This pedagogy is vital. It eschews reductive assessment practices and grading for the sake of competition. It stands opposed to pre-determined learning outcomes and welcomes incidental, unexpected developments that we call learning. It is impassioned and joyous and nerdy. It refuses to measure what is not legitimately measurable. It does not make objects from subjects. It pushes back against any policy that seeks to silence, falsify, or diminish. Failure is critical, as is self-reflection; it loves these processes—it thrives on them.

Moving pedagogy from philosophy to praxis is always a challenge, but the how and the what tend to become visible once I articulate the why and the for/by/about whom. This attempt at a pedagogy of resistance isn't new for me, and yet no matter how pedagogically disruptive I think I am, there's usually further to go. This moment in our history is calling for a full-scale radical overhaul of our systems. It's asking us to reimagine the centers: of pedagogies, curricula, courses, methodologies, and individual lessons inside individual classes. It's asking us to consider who is there, who is not, and why. Perhaps most importantly, it's asking us to consider why not, and why not now.

Critical Citizenship for Critical Times

Maha Bali / مها بالي

Because I have been studying critical thinking for over six years now and live in Egypt, the situation here continues to surprise me. The violence in late 2013 left me stunned. It led me to reflect on critical thinking, citizenship, and what contribution education might make to Egypt's future.

My research (Bali, 2013) has shown that Egyptian high school education makes it difficult for students to question their professors' authority and does not give them confidence to participate critically in classroom discussions. But these same students are more willing to question local and foreign media. Some of them are even willing to question religious authorities.

Despite the educational system that stresses memorization and discourages questioning and creativity, people in Egypt, with many different educational backgrounds, displayed skepticism of the Mubarak regime. While it seems a long time ago now, and much has happened since then, the overthrow of Mubarak was revealing. Despite years of repression, Egyptian youth managed to discern that they needed to get rid of the Mubarak regime. Then they did. It was, and still is, an impressive feat.

Advocacy is considered one of the highest forms of engaged citizenship, and Egyptians have shown they excel at it. However, everything that has come after that uprising, and especially the events of 2013's summer, leave me feeling that Egyptian notions of citizenship are missing something important. Advocacy on the street succeeds in toppling regimes: first Mubarak's, then Morsi's. But that kind of citizenship, based on opposition, seems unable to change tactics and work towards reconciliation and reconstruction. It just recreates the protest cycle over and over again. The escalations of violence surrounding the 3 July 2013 coup d'état further complicate chances for reconciliation.

How much of this failure is due to uncritical citizenry responding to sensationalist media, and how much to factors beyond individuals' agency and control, I don't know. But I believe that higher education has a crucial part to play in preparing today's youth for Egypt's future, including promoting awareness of factors that restrict one's agency to act. I focus on higher education to suggest short-term solutions. Its role extends beyond simply educating enrolled students into community outreach. Long-term, of course, change needs to start in schools.

20.1 Critical Thinking in Higher Education

If promoting citizenship is an overarching goal of higher education, universities need to go beyond just promoting critical thinking (a form of education already in short supply) and community service to focus on developing "critical citizenship." While not necessarily a new concept, the term could help us refocus on what form of education is needed. After years of studying critical thinking, I believe our understanding of critical thinking needs to be contextualized. I work at the American University in Cairo (AUC), and the commonly adopted version of critical thinking here is North American, which includes reflective skepticism to inform decision making. Critical thinking is understood as consisting of a set of skills (such as evaluating evidence, uncovering hidden assumptions, and logically supporting one's argument) and dispositions (such as inquisitiveness and open-mindedness).

Worldwide, it is questionable how far college can develop critical thinking in students who don't already have it. But even this kind of traditional criticality has failed on two fronts. First, most analyses of the Egyptian situation continue to be based on conspiracy theories to explain multiple conflicting realities, with little attention paid to evaluating evidence. Indeed, sometimes there just isn't enough evidence—or even a *search* for evidence. Second, this approach does not prepare citizens to act upon their criticism.

Such action, or "critical citizenship" can benefit from two alternative conceptions of critical thinking.

The first conception borrows from the critical pedagogy movement originating in the work of Paulo Freire (2014). Here, the end goal of critical thinking is to challenge the status quo in order to achieve social justice, collectively raising consciousness of conditions promoting oppression in order to achieve liberation. It is a form of critical thinking that promotes praxis—reflective action based on knowledge, rather than mere activism (which we saw much of in Egypt from 2011 to 2013) or speech and dialogue unaccompanied by action (which we have been seeing for a longer time). It is not mere skepticism about separate facts; it is value-driven and historically situated questioning of power structures that lie beneath the surface. This kind of thinking may be easier to adopt when teaching social sciences and humanities; more complex to include in the study of professions, such as business; and even more difficult in the study of sciences. But it is not impossible. For example, engineering courses can infuse elements of the social, economic, and ethical impact of engineering practices.

The second conception of critical thinking comes from a feminist understanding of critical thinking, based on *Women's Ways of Knowing* by Mary Field Belenky et al. (1986). Their research indicated that women (and some men) tend to prefer more communal and less confrontational ways of learning, rather than the pedagogies usually associated with critical thinking such as debating. This preference to understand the view of "the other" before critiquing it resonates with the philological hermeneutics (understanding a text from the author's viewpoint before critiquing it) of Edward Said (2004). This approach widens one's worldview and also involves elements of empathy missing from the traditional understanding of critical thinking, which prioritizes logic and rationality.

20.2 Critical Thinking in Politics

Political reactions in Egypt seem to me to fall on one of two sides: either complete skepticism regardless of evidence (sometimes even creating fictitious evidence); or complete and blind trust (as in the 26 July 2013 rallies in response to General Al-Sisi's speech). There has also been widespread lack of empathy for how the ouster of Morsi would affect his numerous supporters. The way Egyptians keep dividing themselves, and doing so with passion, makes possibilities for future reconciliation and a pluralistic society difficult, if not impossible.

Egyptians need to develop their own notion of critical citizenship that does not simply adopt ideas from others, but is dialogically and reflectively developed, and responsive to contextual changes, considering issues of social justice and empathy needed in Egypt today. While most academics I know do consider universities agents of social justice, and do themselves have empathetic and social-justice orientations, I believe this does not always reach students, when our focus is to develop a traditional critical thinker. My research (Bali, 2013) has found three pedagogies that can help infuse elements of empathy, social justice, and action in our teaching. The first is apolitical civic engagement via grassroots community service, which research (Assad & Barsoum, 2007; El-Taraboulsi, 2011; Mercy Corps, 2012; Shehata, 2008; Westheimer & Kahne, 1998) has shown promotes adult political engagement. Another is simulated political engagement such as Model United Nations (also Model Egyptian Political Parties suggested by an AUC professor), to explore solutions in a safe environment. The third is intercultural dialogue to widen empathetic understanding of diverse worldviews.

Higher education's role, as I see it, is to help society reflect beyond activism and resistance, necessary and important as they are. There is a need to develop critical citizens capable of negotiating multiple conflicting interests in a process of creatively co-constructing a better future.

It's All About Class: Interrogating the Digital Divide

Lee Skallerup Bessette

I live and work in one of America's poorest regions, Appalachia—specifically eastern Kentucky. Businesses and municipalities don't have a strong web presence (if any at all), Google Maps is essentially useless for getting anywhere, and the social network is still, largely, the local Churches and quilting bees. Howard Rheingold (2012), in his book *Net Smart*, writes about how it is possible now to ask a question and get an answer on your phone anywhere. I hasten to add, as long as it's not *here*, where even cell phone coverage is spotty at best.

Many of my students are part of the "new" digital divide (Crawford, 2011), with limited access to both technology and high-speed Internet. Even students with access to higher-speed cellular service may not be able to afford a data plan or the accompanying smartphone to take full advantage of it. But I think the biggest issue with this new digital divide is not that poorer students are misusing whatever technology they do have (Richtel, 2012), it's that they are not misusing it *enough*. And neither are we.

I have no doubt that many of my students are simply using technology in the same way they "use" television (and even before that, radio): passively consuming content. Technology, for most of the 20th century, has made it prohibitively expensive and thus difficult for people to create and mediate their own content. Just about anyone could write and start a small magazine, pamphlet, or newspaper (assuming rising literacy rates); increasingly, laws prohibited simply starting up one's own radio or TV station. Not that it didn't happen, but, as the classic 90s film *Pump Up the Volume* (Moyle, 1990) shows us, it came with a price. You could speak truth to power, as long as it wasn't on government-controlled bandwidth.

We have, however, come full circle, and once again, to use some dangerous terminology, can somewhat control the means of production. We can write, speak, publish, create, hack, and *play*. But for many, I think, these means of critically interacting with and using technology aren't encouraged because the technology itself is still so expensive. While the cost of owning a computer has generally decreased, it is still a huge financial commitment for many individuals and families. Add to that the cost of what is believed to be required software, Internet access, etc., it is easy to see why students are discouraged from *playing* with their technology—"don't touch that, it's not a toy!" But computers are toys as much as they are tools. Toys, of course, conjure up images of fundamentally trivial games and inconsequential objects; as the saying goes, however, for kids, playing is serious business.

In addition to technology, time, too, is money. There is little time or mental energy for an individual or family trying to make ends meet to just sit and play with their technology. Failure, as well, is more expensive, because if something breaks, there is no time or money to fix it. There are also few resources in the schools to help foster this sense of play and experimentation. In this era of high-stakes testing, suggesting to schools that are "failing" that perhaps what they need is less structured time and more time to play and experiment (particularly with technology) is unthinkable. Once again, the fear of failure, of breaking something, is too great. Firewalls are erected; computers and software are used for drill and kill exercises, if at all; strict rules and guidelines developed and enforced, and tech just becomes one more tool that imposes the banking concept of education on students.

I have been trying to get my students in freshman writing classes to blog, use Twitter, and to *play* with the technology that is available to them. I have always been met with great resistance. For them, Twitter is a waste of time, blogging is just an essay in another form, tech is a tool they have been taught to fear. This is not to say that they don't know how to play, to create, to experiment. One of the reasons they disdain the technology is because many of them don't see how it will help them get a job in their low-tech worlds;

better to know how to hunt, grow gardens, slaughter cows, sew quilts, fish, forage, weld, etc. I am constantly in awe of all they *do* know how to do, versus what I (unfortunately) think they *should* know how to do. I've had students in my class who are computer science majors who only recently bought their first computer; contrast that with what elite engineering school Harvey Mudd requires all first-year students to take: a class where they build an old-fashioned tool the old-fashioned way. This is a real digital and class divide.

This is not to say that the students shouldn't know how to critically engage and use these tools, but that I need to do a better job of bridging, between teacher and student, our dual class divide. But it is becoming increasingly difficult to do so, which reflects the increasing class divide within the university itself. In my own case, English shares a building with the College of Business. Our rooms have blackboards and badly out-of-date whiteboards and computers that only work about half the time, while the business classrooms have new whiteboards and up-to-date classroom technology. Recently, the English department had our two computer labs, often used as classroom space, taken away (the rooms are still there, but the computers are gone). Business still has all of theirs. The message from the university is clear: For English, tech doesn't or shouldn't matter. Focus instead on what you are supposed to be doing: teaching writing. I don't teach naked because of some pedagogical point, I do it because I have to.

But, you might say, a great teacher can find a way to overcome these limitations! This brings up another class issue: tenure-track versus non-tenure-track faculty and the increasing standardization of curriculum, particularly in the writing classroom. I am off the tenure-track, teaching writing according to an increasingly limited script. How much room or freedom do we have to play ourselves, to integrate technology and digital pedagogy in our classroom? How much time do we have to figure out the best ways to help our students learn and engage? What are the incentives? What are the punishments for failure? Low-paid, under-appreciated, and exploited, we are expected, in 15 weeks, to create college-ready writers out of students who don't initially know how to attach a file to an email.

These are the digital divides that worry me, that discourage and prevent many teachers from embracing a more hybrid form of teaching. Both students and teachers need the support and encouragement to play, to have the time and fearlessness to use and "misuse" tech.

Pedagogy & the Logic of Platforms

Chris Gilliard

> Computers are just as oppressive as before, but smaller and cheaper and more widespread. Now you can be oppressed by computers in your living room. (Nelson, 1987, p. 20)

In his initial New Horizons column in EDUCAUSE *Review,* Mike Caulfield (2017) asked, "Can higher education save the Web?" I was intrigued by this question because I often say to my students that the Web is broken and that the ideal thing to do (although quite unrealistic) would be to tear it down and start from scratch.

I call the Web "broken" because its primary architecture is based on what Harvard Business School Professor Shoshana Zuboff (2015) calls "surveillance capitalism," a "form of information capitalism [that] aims to predict and modify human behavior as a means to produce revenue and market control" (75). Web 2.0—the Web of platforms, personalization, clickbait, and filter bubbles—is the only Web most students know. That Web exists by extracting individuals' data through persistent surveillance, data mining, tracking, and browser fingerprinting (Goodin, 2017) and then seeking new and "innovative" ways to monetize that data. As platforms and advertisers seek to perfect these strategies, colleges and universities rush to mimic those strategies in order to improve retention (S. Brown, 2017).

That said, I admit it might be useful to search for a more suitable term than "broken". The Web is not broken in this regard: A Web based on surveillance, personalization, and monetization works perfectly well for particular constituencies, but it doesn't work quite as well for persons of color, lower-income students, and people who have been walled off from information or opportunities because of the ways they are categorized according to opaque algorithms.

My students and I frame the realities of the current Web in the context of digital redlining, which provides the basis for understanding how and why the Web works the way it does and for whom. The concept of digital redlining springs from an understanding of the historical policy of redlining: "The practice of denying or limiting financial services to certain neighborhoods based on racial or ethnic composition without regard to the residents' qualifications or creditworthiness. The term 'redlining' refers to the practice of using a red line on a map to delineate the area where financial institutions would not invest" (The Fair Housing Center of Greater Boston, n.d.).

In the United States, redlining began informally but was institutionalized in the National Housing Act of 1934. At the behest of the Federal Home Loan Bank Board, the Home Owners Loan Corporation created maps for America's largest cities and color-coded the areas where loans would be differentially available. The difference among these areas was race.

Digital redlining is the modern equivalent of this historical form of societal division; it is the creation and maintenance of technological policies, practices, pedagogy, and investment decisions that enforce class boundaries and discriminate against specific groups. The *digital divide* is a noun; it is the consequence of many forces. In contrast, *digital redlining* is a verb, the "doing" of difference, a "doing" whose consequences reinforce existing class structures. In one era, redlining created differences in physical access to schools, libraries, and home ownership. In my classes, we work to recognize how digital redlining is integrated into technologies, and especially education technologies, and is producing similar kinds of discriminatory results.

We might think about digital redlining as the process by which different schools get differential journal access. If one of the problems of the Web as we know it now is access to quality information, digital redlining is the process by which so much of that quality information is locked by paywalls that prevent students (and learners of all kinds) from accessing that information. We might think about digital redlining as the level of surveillance (in the form of analytics that predict grades or programs that suggest majors to students). We also might think about digital redlining to the degree that students who perform Google searches get certain information based on the type of machine they are using or get served ads for high-interest loans based on their digital profile—a practice Google now bans (Graff, 2016).

It's essential to note that the personalized nature of the Web often dictates what kind of information students get both inside and outside the classroom. A Data & Society Research Institute study makes this clear: "In an age of smartphones and social media, young people don't follow the news as much as it follows them. News consumption is often a byproduct of spending time on social media platforms. When it comes to getting news content, Facebook, Twitter, Instagram and native apps like the Apple News app are currently the most common places where the teens and young adults ... encounter news" (Madden et al., 2017, p. 20).

Students are often surprised (and even angered) to learn the degree to which they are digitally redlined, surveilled, and profiled on the Web and to find out that educational systems are looking to replicate many of those worst practices in the name of "efficiency," "engagement," or "improved outcomes." Students don't know any other Web—or, for that matter, have any notion of a Web that would be different from the one we have now. Many teachers have at least heard about a Web that didn't spy on users, a Web that was (theoretically at least) about connecting not through platforms (Friend & Gilliard, 2018) but through interfaces where individuals had a significant amount of choice in saying how the Web looked and what was shared. A big part of the teaching that I do is to tell students, "it's not supposed to be like this," or, "it doesn't have to be like this." The web is fraught with recommender engines and analytics. Colleges and universities buy information on prospective students, and institutions profile students through social media accounts (Singer, 2013). Prospective employers do the same. When students find out about microtargeting, social media "filter bubbles," surveillance capitalism, facial recognition, and black-box algorithms making decisions about their future—and learn that because so much targeting is based on economics and race, it will disproportionately affect them—their concept of what the Web is changes.

Another aspect of my teaching is rethinking the notion of "consent." It's important to ask: What would the Web look like if surveillance capitalism, information asymmetry, and digital redlining were not at the root of most of what students do online? We don't know the answer. But if higher education is to "save the Web," we need to let students envision that something else is possible, and we need to enact those practices in classrooms. To do that, we need to understand "consent" to mean more than "click here if you agree to these terms."

I often wonder if it's possible to have this discussion without engaging in a deep and ahistorical practice of nostalgia. Telling students about the "good old days" of hand-coding and dial-up Internet access probably isn't the best way to spend classroom time. However, when we use the Web now, when we use it with students, and when we ask students to engage online, we must always ask: What are we signing them up for? (Ultimately, we must get them to ask that question themselves and take it with them.) Here the term "consent," often overused and misunderstood, needs to be foregrounded in the idea that we must do all we can to explore the reality that students are entering into an asymmetrical relationship with platforms.

While we can do our best to inform students, the black box nature of the Web means that we can never definitively say to them: "This is what you are going to be a part of." The fact that the Web functions the way it does is illustrative of the tremendously powerful economic forces that structure it. Technology platforms (e.g., Facebook and Twitter) and education technologies (e.g., the learning management system) exist to capture and monetize data. Using higher education to "save the Web" means leveraging the classroom to make visible the effects of surveillance capitalism. It means more clearly defining and empowering the notion of consent. Most of all, it means envisioning, with students, new ways to exist online.

(dis)Owning Tech: Ensuring Value and Agency at the Moment of Interface

Timothy R. Amidon

Education is big business. In the U.S., over 5% of gross domestic product is earmarked for education (U.N. Human Development Report Office, 2013). Student debt in the U.S. is estimated to be over $1.2 trillion (Chopra, 2013). The educational technology market is worth over $8 billion (Chen, 2015). Certainly, a great deal of economic value circulates within and through educational systems. However, schools and colleges also create forms of social, cultural, humanist, and civic value. In a *Hybrid Pedagogy* CFP, Chris Friend (2015) challenges educators to interrogate "[how] critical digital pedagogy [can] add to or ensure the value of an education." Foregrounding the critical role that autonomy plays within learning, Chris gestures tacitly toward the decreasing level of agency that those most directly involved in learning have in defining the processes and purposes of education on their own terms: "Teachers must choose to create classes and schools wherein students actively create their learning environments and control their own progress." My interest was piqued.

Like Chris, I share a concern for critical hybrid pedagogy and view the purpose of education as "human enrichment and increased consciousness" (Friend, 2015). As an intellectual-property (IP) scholar, an associate editor with *Kairos: A Journal of Rhetoric, Technology, and Pedagogy*; a member and former chair of the IP Caucus of the *Conference on College Composition and Communication*; and an assistant professor at Colorado State University (CSU), I've spent a good deal of my early career learning how ownership impacts the type of intellectual property that is created and consumed within schools and universities. Through this work, I've come to realize that decisions regarding ownership in educational systems are always decisions about:

1. who will (and will not) control and define the learning process,
2. who will (and will not) profit from the ways that learning processes are enacted,
3. who will (and will not) have access to science and scholarship and the infrastructure necessary for creating it,
4. who will (and will not) participate in the design of curriculum and assessment and learning spaces,
5. who will (and will not) profit from the benefits of science and artistry, and
6. who will (and will not) have opportunities to attend schools and colleges.

Thus, while I strive to enact a critical pedagogy that is built around the type of self-actualization that enables students to realize the civic and humanistic aims of education, I often struggle to achieve the level of autonomy that seems necessary for this work because authority over, and ownership of, education has been distributed to a wide variety of stakeholders—many of whom seem to frame the purposes and value of education in purely economic terms.

Within my home institution, for example, we might sketch a network of stakeholders that consists of government-sector actors such as elected officials and Colorado Department of Higher Education; public-sector actors such as taxpayers, current and potential students, alumni donors; and private-sector actors such as local and global businesses that employ CSU graduates, student-loan financiers like Sallie Mae, designers of tests like Advanced Placement and CLEP, learning management systems like Instructure's Canvas, educational technologies or iParadigms' Turnitin, as well as databases like ProQuest and EBSCOhost. A network perspective not only lays bare the various stakeholders with a vested social, economic, and political interest in what happens within schools and colleges, but also the ways agency for what happens within classrooms at my institution extends beyond the students and educators charged with constructing learning. Cultivating environments for agentive learning occurs within educational systems where ownership is increasingly distributed, so students and educators who see the value of humanistic, civic education must do so while negotiating the multiple, competing aims promulgated by other stakeholders.

Consequently, it's productive to not only think of schools and colleges as sites of learning, but also as marketplaces where goods, knowledge, and services are consumed and produced. It's reasonable for private-, public-, and government-sector actors to be motivated in different ways. However, there is a fundamental flaw when education systems—however distributed they might be—place the aims of individual profit and privilege before the humanist and civic aims of education. Value is created in educational systems that *equitably* expand the wealth of human knowledge through science and artistry; produce workforces that can participate within economic systems that create the knowledge, goods, and services that societies consume; and prepare students "to participate fully and meaningfully" (Selber, 2004) as citizens of a democracy. Unfortunately, the civic and humanist aims of education—those of greatest consequence to us collectively—seem to be those which are most readily obscured by and subordinated within current political discourse surrounding education in the U.S. Too often, this discourse reduces (or seeks to reduce) our schools to businesses and education to, as Chris Friend (2015) framed it, "a sales transaction."

Powerful examples of this discourse can be found, for instance, within the destructive "Dear Student," "Work Harder," and "Broken System" narratives (McCalmont, 2015; Mehta, 2013; Patton, 2015) that pit students, teachers, and publics against one another. These narratives are so problematic because they offer scapegoats that draw attention away from the "ongoing erosion of state support" (Colorado Commission on Higher Education, 2010) which has occasioned the legitimate anxiety students and teachers feel about their agency over how learning happens in classrooms. As operational costs of running and maintaining universities and schools have grown, administrators have turned toward selling off pieces of our educational systems (Bok, 2004). Some argue that "schools have slowly and steadily improved" (J. Schneider, 2016), but for many divestment from public education is at the heart of very real issues (Zernike, 2016) ranging from the reliance on contingent labor to staff courses and the swelling transfer of the costs of education to students in the form of student debt, to the role that third-party testing and assessment services play within with our schools and the rush to automate instruction through the use of educational technology. Of course, all of these issues deserve sustained critical attention, but this article is particularly concerned with the ways that uncritical adoption of educational technologies adversely impacts the autonomy of students and teachers within the shared enterprise of learning.

23.1 Interfaces and Agency

Across the two decades I've spent in higher education, I've watched certain technologies become a central component of how we enact education. Cell phones and email didn't meaningfully exist within my life as an undergraduate. Now, as a faculty member, I am tethered to Gmail, Google Drive, Canvas, and Twitter for what seems like hundreds of hours a week. It often seems like a struggle to construct agency amidst the many technological interfaces (many that I haven't autonomously chosen for myself) that I encounter in both my scholarly and personal life. It is precisely this struggle that has drawn me to computers and writing, a field that has long attended to the nexus of interface, agency, and literacy. What I have to say about interface is deeply influenced by Anne Frances Wysocki and Julia Jasken (2004), Cindy and Dickie Selfe (1994), Stuart Selber (2004), Jeff Grabill (2003), Ellen Cushman (2013), Doug Walls, Scott Schopierary, and Dànielle Nicole DeVoss (2009), DeVoss et al. (2005), and W. Michelle Simmons and Grabill (2007) whose scholarship reveals how cultural, legal, political, social, and economic values are built into the interfaces.

Indeed, this body of scholarship might be read as examples of the different ways agency and power intersect within distinct contexts. Wysocki and Jasken (2004), for instance, observe that "interfaces are about the relations we construct with each other." DePew and Lettner-Rust (2009) argue that "interfaces...mediate other power relations between instructors and students." Similarly, Sean Michael Morris and Jesse Stommel (2012) caution that educators might be uncritically relinquishing control of pedagogy to educational technologies such as LMSs and encouraged educators to shift their stances to ensure "pedagogy [is] driving functionality." However, a great deal of the political work that influences the shape of these interactions occurs beyond the screens. As Selfe and Selfe (1994) contend, interfaces are not merely neutral sites of exchange but locations which reflect dominant power structures:

> Borders are at least partly constructed along ideological axes that represent dominant tendencies in our culture...borders evident in computer interfaces can be mapped as complex political landscapes...borders can serve to prevent the circulation of in-

dividuals for political purposes, and...teachers and students can learn to see and alter such borders in productive ways. (481–82)

What I find most inspiring about this scholarship is that it not only understands that interfaces are built to reflect values that circulate in social, political, and economic spheres, but also that interfaces are designed and built by people who have the agency to change them. Educational technologies, as interfaces, offer students and educators opportunities to discover and enact agency through strategic rhetorical action. Yet, realizing this agency is complex work because "participat[ing] fully and meaningfully in [the] technological activities" (Selber, 2004) that comprise so many aspects of our social, civic, and professional lives requires an increasingly sophisticated array of multiliteracies. Indeed, Selber's work on computers and literacy suggests that agency is realized through a blended repertoire of functional (ways of doing), critical (ways of knowing), and rhetorical literacies (ways of reflexively, ethically, and agentively combining functional ways of doing with critical ways of knowing). Thus, agentive action not only requires a deep understanding of the value assumptions undergirding the technological interfaces we encounter, but also the technological and rhetorical prowess necessary for enacting changes to the ways we connect and relate at the moment of interface.

23.2 Values and Interfaces: What I learned from using Eli Review

To be clear, I am not against buying and selling: this isn't an invective about whether or not commercial models are to be lauded or shamed or whether they have a place in education. Following Lawrence Lessig (2007), I see value in both sharing and competitive economies, because they have different motivational structures that incentivize different types of work and projects. For instance, there are commercial actors within rhetoric and composition like Bedford St. Martin's and Eli Review who add value to our field by employing scholars who participate within and enrich our disciplinary conversations, who listen and act transparently, and who make products and technologies that reflect pedagogies our discipline values as sound and ethical. Eli Review, for instance, was designed by writing teachers who understand that writing is best taught as an iterative and social process, and have built an interface that facilitates peer- and teacher-feedback within learning. They regularly host free teacher development workshops. Moreover, when I wanted to improve how I was using the tool in my own composition courses, Eli partnered me with their Director of Professional Development, Melissa Meeks, who helped me redesign the prompts from the ground up and regularly met with me to check in on how things were going in the course. For me, this level of involvement was uncommon for an educational technology company. It suggested that Eli wasn't simply designed as a product for purchase, but as a technology with support mechanisms built into both their approach and interface. It was a technology that came with human support so that I could ensure that the tool helped me to scaffold the peer-to-peer and instructor-to-peer interactions that offer students generative experiences writing in our course.

The challenge is that technology designers aren't always motivated by the same values or the right balance of values. Lawrence Lessig (2004) argues that we must strike an ethical and sustainable balance with how we award ownership through copyright because extending monopolistic control of ideas into perpetuity is neither good for innovation nor consistent with the aims democracy. Yet, a great deal of research and scholarly knowledge produced within publicly funded universities is published within journals owned by commercial publishers such as Elsevier, which makes over billion dollars a year in profit (MIT Libraries, 2020). For Lessig, U.S. copyright wasn't designed as a mechanism to protect unfettered profiteering, but rather a legal protection that placed paramount value on the future accessibility and sustainability of scientific and artistic content because of the central role such knowledge plays within a democracy.

Thus, it's not surprising that Jeff Grabill, one of the creators of Eli Review, posits that students would be well served by critical pedagogues who attend to the values that are instantiated within interfaces. Pointing toward the increasing role technologies play in learning environments, Grabill carves a careful distinction "between educational technologies (or technologies for) and learning technologies (or technologies with):"

> Technologies *for* automate teacher work. And, if we had more time, I'd tell you about my decade [of experience] in the educational venture capital world where they are out to replace you. [They] replace the teacher, focus on testing, focus on summative feedback, de-professionalize teacher work, and [they're] free.
>
> I want to dwell for a minute on *free*. Academic humanists might be the last people in the world who believe there is something called *free*. There is no such thing as free. You're paying for [your technology]—you might be paying for it by making your students give up their personal data, or by giving up your own data, or you may be giving up technical or learning support. But you're paying for it, and one of the most insidious moves in ed-

ucational technology in K-12 is schools penchant for *free* on the surface, which costs them dearly downstream, particularly in the toll it takes on the lives of teachers and lives of students. (Eli Review, 2016)

Through the practices like mentoring, partnering, supporting, and being responsive, Eli Review has built an interface that *informates* what teachers and students do. It's supplemental and it is designed to scaffold and enrich the agency that students and teachers have over the process and products of their learning.

Conversely, there are "educational technologies" designed to *automate* the work of teachers. For instance, iParadigms' Turnitin employs a rhetoric of fear to turn educators away from, as Rebecca Moore Howard (2007) puts it, "pedagogy that joins teachers and students in the educational enterprise [by choosing] ... a machine that will separate them," but also leaches intellectual property students create within educational systems, only to sell it back to schools. Unfortunately, plagiarism detection software (PDS) like Turnitin has been so widely (and uncritically) adopted that members of the Conference on College Composition and Communication (CCCC), "the world's largest professional organization for researching and teaching composition," passed for a formal resolution warning colleagues about the ways that PDSs "compromise academic integrity" during their 2013 meeting based on a position statement that the CCCC-Intellectual Property Caucus drafted (CCCC-IP Caucus, 2006).

23.3 Values and Interfaces: What I learned about the Relationship between ProQuest and Turnitin

A well-established body of scholarship within rhetoric and composition explores the ways PDSs violate student intellectual property and adversely position students and teachers (Price, 2002). Moreover, PDSs just aren't capable of drawing nuanced distinctions between *actual* plagiarism and the type of patchwriting that commonly reflects a writer's learning how to master academic literacies like paraphrasing and using citation systems (Howard, 1999). Consequently, I was surprised when I learned through a series of Twitter conversations (Stommel, 2015) that a scholarly company I respected and trusted, ProQuest, had an existing relationship whereby they had provided Turnitin access to content in their databases. For a rich discussion of the history and arguments surrounding PDSs, see "Turn It Down, Don't Turn It In: Resisting Plagiarism Detection Services by Teaching about Plagiarism Rhetorically" by Stephanie Vie (2013). In fact, I had recently provided Pro-

Quest with access to my dissertation. As a member of the CCCC-IP Caucus, I am ideologically opposed to Turnitin, and it upset me that this company might have access to intellectual property I created. Still what *really upset me* was that relationship wasn't made transparent when I interfaced with ProQuest's Electronic Theses and Dissertations (ETD) database. I understood that the Library of Congress and the University of Rhode Island, institutions I trust, value ProQuest because they make an important contribution to the progress of science and knowledge by curating their ETD repository. And, because I trusted URI and the Library of Congress to bestow that responsibility with ProQuest, I, in turn, trusted ProQuest.

I also understood, following the language in the University Agreement statement, that I retained "the rights in copyright for theses and dissertations produced as a part of a University degree" but that I had, "as a condition of the award of any degree, grant[ed] a royalty-free license or permission to the University and any outside sponsor, if appropriate, to reproduce, publicly display on a non-commercial basis, copies of ... student dissertations." I have been revising portions of my dissertation for publication as articles and chapters, so I utilized the embargo option in order to protect my scholarship from the first publication clause. But, after learning that ProQuest had potentially been sharing my intellectual property with a company that I disavow, I wanted to know more about how ProQuest grants Turnitin access to content in their database. I wrote to ProQuest and exchanged correspondence with a number or representatives. During my initial exchange, I learned that ProQuest considers Turnitin a third party search engine, which I would argue is, at best, a disingenuous way of representing a PDS. They informed me that:

> Our records indicate that you did elect to allow third party search engine access however you do have an embargo on your work since the time of submission therefore your work would not have been supplied to Turnitin.com by ProQuest.

Indeed, I did elect this understanding; ProQuest's interface suggested that doing so enables "major search engines (e.g., Google, Yahoo) to discover my work through ProQuest." As I continued to correspond with various representatives, I was informed that ProQuest "[had] not distributed [my] manuscript to any 3rd party indexer." Since that time, ProQuest has authored more substantive clarifying language regarding their partnership with iParadigm/Turnitin/iThenticate (ProQuest, no date).

Yet these statements are not built into the actual interface students navigate to make agentive decisions regarding how they share their content. Moreover, some institutions unilaterally compel students to submit work to

Turnitin/iThenticate as a condition of granting a degree rather than trusting students and their advisors to produce and ensure the veracity and originality of scholarship. I am thankful that the graduate school at URI did not take this stance while I was a student there. Still, I continue to be concerned about this tweet that Turnitin posted on 8 March 2012: "RT @ithenticate: 300,000 dissertations and theses from @ProQuest added to @TurnItIn and @iThenticate database #highered"

As this tweet and a press release from 12 March 2012 seem to suggest (Lawrence, 2012), it appears that ProQuest has offered Turnitin access to their database. So which is it? Were the authors who uploaded their IP to the ProQuest repository prior to the formation of the ProQuest-Turnitin partnership consulted about whether they assented to share their IP with this third party? Has Turnitin/iParadigms/iThenticate indexed works which they should not have had access to? Interfacing with this technology makes me curious about whether student-authors have agency over who accesses their work and how others might profit from labor that takes place in publicly funded institutions. Examining ProQuest in relationship to the Vanderhye ruling (United States Court of Appeals, 4th Circut, 2009), which held iParadigms' indexing was legal under fair-use provisions of copyright law, there are notable distinctions to be drawn. Specifically, the ProQuest interface does not make it transparent that copyright holders may be sharing their IP with iParadigms, nor have copyright holders entered into the same legal agreement as students in the Vanderhye cases had when they clicked "I agree" and actively uploaded their content, in turn, providing the indexical database access to their intellectual property. Even if "fair use doctrine protects the transformative uses of content, such as indexing...," (Macgillivray, 2009) companies like Google offer insights into how agency might be returned to creators who wish to uphold their moral rights:

> Even though the Copyright Act does not grant a copyright owner a veto over such uses, it is our policy to allow any rightsholder...to remove their content from our index....

This policy statement reflects that Google's stance is that they seek to affirm, ultimately, the agency of creators. By adopting a shift in stance towards how PDS are implemented, educators could work toward affirming the agency of student-creators. Rather than using coercive power to force students to submit to PDSs, educators could offer the service to students who wish to use it. Similarly, PDSs could similarly follow Google's approach by allowing creators to remove content that has been indexed. Offering opt-in and opt-out mechanisms to student-creators demonstrates respect for their decision-making capacity as humans. If the purpose of education is to serve humanist and civic aims, critical pedagogues must work to ensure that the structures through which it is enacted (institutions, classrooms, technologies, databases) should empower the people those structures were designed to serve. Critical pedagogy is about ensuring that learning is grounded in an ethic that serves humanist aims and enriches our communal, civic well being. It creates and protects places where hybridic ways of being, knowing, and doing are possible. Through tactical action, critical pedagogues might adapt how they relate with students during moments of interface, but they must also attend to the ways that pedagogical interface is distributed to educational technologies. Critical pedagogues must work to ensure that the technologies that serve us, serve our aims.

(Higher) Education as Bulwark of Uselessness

Luca Morini

In 2014,[1] halfway through the twisting path that was my doctoral course, I found myself in Finland, at the Critical Evaluation of Game Studies Seminar, where, above all the "big names" in the field of Game Studies who spoke there (among which were Espen Aarseth, Jesper Juul, and Frans Mäyrä), one thing was indelibly imprinted in my memory: Canadian sociologist Bart Simon's characterisation of Game Studies as a true, undeniable "bulwark of uselessness", a field of thought that can work in resistance to all appeals to productivity and efficiency. Because really what can be more frivolous, in "productive" common sense, than spending a couple of days discussing the philosophy of play and games?

---

[1] This article, originally published in July 2016, was a "twin essay" to "Play as Bulwark of Uselessness" (Morini, 2016), published on *First Person Scholar*, a journal of 21st Century media cultures. The twin essays were composed simultaneously, as a playful experiment in academic writing: Beside some words or remarks, the first and last paragraphs are indeed just the same, highlighting the deep, if most often distorted—as discussed in both essays— link between play and learning.

As someone with a preference to play supportive and protective roles in online games (or, to follow gaming jargon, a "tank"), always relishing the role of defending my teammates in our small, unnecessary virtual struggles, I found the image stuck strongly.

As I continued climbing toward the completion of my Ph.D. in Education and Communication, largely by playing and making games within communities of amateur game designers, I came to think that this powerful image, that of "the bulwark of uselessness", could be a conceptualisation relevant to all cultural endeavours, in their conflicting relationship with narrowly utilitarian economic forces. I reflected on how, in the current historical-cultural moment, this bulwark finds itself attacked (D. Hill & Kumar, 2012) in its last public expressions, that is, the spaces of institutionalised education in general and university in particular, and that the fall of this bulwark, its full exclusion by public spaces, cultures and discourse, would be nothing short of catastrophic.

As a living example of these utilitaristic attacks, having soon after my graduation obtained a research position at an institution aimed at promoting change in higher education, Coventry University's Disruptive Media Learning Lab, I found myself involved in a variety of projects that most often, with different degrees of subtlety, involve a particular, compliance-oriented mode of learning: gamification (Deterding, 2014; Watson, 2013). At the time, it meant for me seeing games and play used as tools, focused on pushing people to address some specific instrumental purpose (luckily just as often noble ones, such as promoting more environmentally sustainable practices) just as all too often education itself seems close to becoming merely a simple tool for economic ends (J. Walsh, 2018).

That was an uncomfortable position, one that I struggled to critically come to terms with as an engaged pedagogist and game scholar. While I participated in a (still ongoing) push towards a more co-creative approach to playfulness and games (Arnab et al., 2018), my strong concerns with utilitarianism were one of the factors that contributed to my move to a different department, Coventry's Institute for Global Learning, where I endeavour to focus more specifically on issues of international and intercultural education—though I still concern myself with the different types of "gaming" that the metrification of education has led to (Csiszar et al., 2020).

24.1 Because, really, what is "uselessness"?

Being part of the above mentioned "bulwark of uselessness" is to push every day against mainstays of 21st Century University discourse—such as the masterful victim blaming that is the modern concept of "employability" (Rooney & Rawlinson, 2016)—which push to expunge imaginative and critical cultural work from the public sight, in favor of an exclusive focus on what is considered immediately "useful": management and productivity. Higher education seems not to be about education anymore, inclusive of cultural, political, epistemological and ethical considerations, but more and more about mere technical, specialised training, going as far as to quantify one's patterning at human relationships as "soft skills", which are usually conveniently aimed at furthering a corporate agenda (Fixsen, 2017). How come employers give credit to supposed "skills" like "adaptability" and "conflict resolution" and not to the likes of "political awareness", "resistance to authority" or "union organising"? I surely learned a lot about the latter in the last few years.

I even heard—and read (Jorre de St Jorre & Oliver, 2018)—fellow academics saying, in complete good faith, that this is what our students want, and we are therefore being democratic in providing them with theory-devoid curricula. This way they can focus on training for concrete, practical stuff that makes up the "real world" and therefore, ultimately, get a better job. And it's not like those teachers who say so necessarily like educational institutions' current "market orientation". They just think it is a fact of reality, something that just is. Remembering how Alfred Korzybski (1958) and Robert Anton Wilson (2012) warned us against using the word "is", it constantly appalls me how close "market" and "reality" end up being in this field of discourse—which also conveniently obscures how, however, structural inequality doesn't care about "employability skills". Just have a look at graduate employment data involving gender and ethnicity, wherever you live.

(Higher) education is useless. This is a point of view we keep hearing more and more in the media. It started with the widespread irony on philosophy majors' job perspectives (Dominus, 2013) and is now expanding to everything aesthetics or theory-related. In later years, this argument possibly reached its peak in 2018 with a book (that I will not cite) by Bryan Caplan in which he, after correctly diagnosing how education falls short of the promise of somehow "solving" inequality, argues that we might as well stop investing in it.

The main point of this essay is however that we should not reject this pervasive rhetoric in its various versions, and this because we can't reclaim utility for our endeavours without implicitly submitting to the tyranny of productivity. We can't claim legitimation using the same criteria of our opponents, or, to quote Audre Lorde (2018), we can't use the master's tool to dismantle the master's house. We need, instead, to embrace this "accusation" to its very end, following the advice of anthropologist and system thinker Gregory Bateson (1979) on confronting paradoxical situations: there is no way out; the only way is through. Please note that there is no irony in my claim to higher education's uselessness, if not the specific choice of word (I could have gone for less provocative alternatives, such as "anti-utilitarianism" or "unproductiveness"). I am being completely straight in claiming that the role and glory of education is that it can be useless, not being bounded by criteria of production and pre-determined purpose.

24.2 Of Uselessness and Dinosaurs

Many philosophers and thinkers have proposed poignant critiques of the discourse of utilitarianism—from Marx to John Paul II, to cover the political spectrum (Brenkert, 1981; Colosi, 2020)—but still, these are just theories, not grounded in "reality" or "facts". Let us then take a shot at this "going through" approach and, for a while, lean not on Philosophy or Pedagogy, but on so-called "hard sciences" themselves, those same, "useful", "productive" STEM subjects that are being rhetorically used to push humanities and criticism outside of the academia.

There is this concept in evolutive biology, *exaptation* (Gould & Vrba, 1982), which describes shifts of function in evolved traits. To make a macroscopic example: Feathers (probably) initially evolved in dinosaurs for thermoregulatory purposes—a warm coat which let this atypical reptiles be more efficient and expand the spaces of their living to northern and southern latitudes. A very useful adaptation, there is no doubt in that.

Some dinosaurs, however, were not content with just being warmed by their feathers and started playing with them. They even showed them off to potential partners, and, for these unwholesome and useless practices, were probably censured as slackers, time-wasters and societal burdens by their elders and bureaucrats (please do remember this essay is a "playful experiment").

Then, one day, maybe by accident, maybe while playfully chasing each other, maybe falling from a tree they were singing some serenade on, one of them realised that this slick coat had interesting, unexplored and unforeseen aerodynamic properties. In a few generations, a whole new dimension of being suddenly opened up, and when unforeseeable catastrophe struck in the form of an ecosystem-changing asteroid, the grandchildren of those shameful slackers were the only ones to survive, in the form of modern birds.

Of course this is a gross simplification of evolutionary dynamics (and of dinosaurs' social structure) for narrative's sake. Still, to quote Professor Ian Malcolm from Jurassic Park (Spielberg, 1993), "life finds a way", always, and my main argument here is that this happens way less painfully and dangerously when it has been allowed to play free. What is now useless can open up whole new worlds tomorrow. And even if it never does, it is beautiful, in that it has the markings of the play of possibility that is life and mind. It should not need to be justified against "productivity" or "learning outcomes" checklists.

24.3   All Work and No Play

This is the purpose of education for me: to allow the mind to play free of purpose itself. (Again echoing Bateson, 1979, I use the word "mind" in its largest definition, inclusive of social and living systems.) Neo-liberal thought, by contrast, pushes us into an eternal, immutable present, a perversion of the very concept of "sustainability" (Mattei, 2014), as in being able to indefinitely perpetuate the same market dynamics—a patent impossibility given their extractive nature. For this titanic task to be accomplished, there is a need to lock out the production of alternative, possible worlds, especially those who are deemed wasteful and unproductive, so that any kind of entertainment which is not closely related with consumption (or even if it is not consumption of things bought, not freely gifted) is labelled as "escapism". Which, as C. S. Lewis (2002) famously quipped, is a preoccupation of jailers.

Following this train of thought, one prominent example of these assaults on uselessness pertains to the close relationship between play, games, and education. As of today the large majority of games and playful practices are still preponderantly expelled by the places of learning—and were, until very recent times, most often reviled and used as scapegoats by the media (Copenhaver, 2019). However, at the very same time learning institutions, influenced by technocratic (and techno-deterministic) stakeholders, spend millions of dollars in "serious games" and "gamification". These are ludic (or para-ludic) practices characterised first and foremost by their "telic character", their purposiveness (Stenros, 2015), the idea that games are fun but only really worth our time if they can also do some "useful work" (Star, 2013).

Even when many of these games or "gameful" practices promote healthy or sustainable practices, on a metacommunicative level they convey another, hidden curriculum, that of surveillance, efficiency, skill, or information delivery and, above all, compliance, the reduction of behavior to the useful and foreseeable (Watson, 2012).

How can we, instead, meta-communicate liberation and possibility?

Again drawing on game cultures, a good inspiration for how we can do this is the rising global movement (exemplified by the diverse likes of Paolo Pedercini, Rosa Carbo-Mascarell, Gonzalo Frasca, Zoe Quinn and Anna Anthropy, among many others) that promotes freely creating and sharing games from extremely limited resources (Anthropy, 2012).

Following their example, we can liberate education by never renouncing the uselessness and playfulness that should characterise true learning (Parker-Rees, 1999), whatever the forms that "play" assumes, be it on a stage; on a musical instrument; with feathers; or with an amateur, purposeless digital game we ourselves designed, developed and shared. Anything goes, as long as it eludes the hegemonic criteria of market and productivity, and preserves the voluntary, joyful character of play (Suits, 2014).

But what are the risks, if we do not nurture, sustain and promote a cultural stance that allows for what Roger Caillois (2001) calls "pure waste"? What could happen if we let this bulwark of uselessness that is education (and higher education in particular) fall?

24.4 The Tyranny of Necessity

Through our current digital ecologies we can play with people we would otherwise never even meet. Even better, we can create spaces for people we have never met to be playful in, and to learn together. We can go beyond the tyranny of proximity and provincialism, beyond the economic tyranny of utility, and even the ontological tyranny of necessity (Suits, 2014).

The key of my argument, however, is that the fall of this "bulwark" under the blows of efficiency and utility would indeed constitute the ultimate ecological catastrophe: Would the spaces where novel ideas can emerge unbound by efficiency or productivity be eliminated, the consequence would be no less than the simultaneous destruction of all non-actual possible worlds, collapsed in the monolithic, eternal present of the capitalist "end of history" famously discussed by Francis Fukuyama (1989) after the fall of the Berlin Wall. That is, until a further catastrophic eco-social collapse necessarily happens, due to the utmost rigidity of such a system, and we go the way of the non-flying dinosaurs. (Please note that I'm not abstractly worried about "the planet". In the immortal words of George Carlin [qtd. in Dadniel, 2007], "The planet is fine. The people are fucked.")

In this context, the most ethical "purpose" of education can therefore be only and exactly to critique purposiveness itself, a critique which, in its praxis, comes in (at least) two flavors:

- To create safe spaces for the emergence of practices and systems which purposes are not known yet, and might never find one.
- To strip existing practices of their current purpose, letting new ones, unbound by current utilitaristic imperatives, emerge.

I want therefore to conclude echoing Henry Giroux (2001) and his remarks against efficiency, seriousness, and technical determinism. My appeal is to the citizens of an insidiously colonised land, spaces no more completely public, but more and more subjected to market forces and imperatives. My appeal is to get involved wherever there is the possibility of critical education through playful subversion, something that, indeed, even our current, colonised learning institutions still allows and provides space for, if often unknowingly and implicitly. See, for a paradigmatic example, the massive cheating and playful boycott practices which characterized Italy's INVALSI standardised school evaluation tests, as a creative resistance to measurement (Millozzi, 2015).

Subvert mere institutionalised training into engaged, playful education because, contrary to the famous saying by Margaret Thatcher (1980), there are, indeed, infinite alternatives. As teachers, educators, and pedagogists, our entire job consists in cultivating these alternatives—these possible worlds—and this is something we can keep doing only by upholding the bulwark of uselessness: legitimating "suspensions of productivity", both your own and others', as "useless" spaces are actually the most evolutive ones, those that can generate alternatives, and resist the instrumental purposes, of our so-called, common-sense "reality".

The Political Power of Play

Adeline Koh

We are accustomed to thinking about play as frivolous. We think of play as something that young children do; play is not serious, it doesn't encourage deep, intellectual thought, it must be set aside as one grows older for quiet, reserved contemplation. Play is fun and pleasurable, the supposed opposite of rigorous education. Yet, Fred Rogers (1995) (better known as Mr. Rogers) is well known for his claim: "Play is often talked about as if it were a relief from serious learning. But for children, play is serious learning. … Play is really the work of childhood" (p. 47). People who work in the arena of higher education have extended this sentiment to grown-up children: Pete Rorabaugh, Sean Michael Morris, and Jesse Stommel (2013) argue that play constitutes a new form of critical inquiry; Cathy N. Davidson (2011) suggests, in *Now You See It*, that game mechanics should be used to reformulate some of the most critical learning goals in education; game designer and evangelist Jane McGonigal (2011) notes that "reality is broken" and that games are the solution to many of our problems—that if we played games as if our lives depended on them (especially collaborative games), we would learn that challenges never stop, and that it is worth risking absolute failure for an epic win. Accordingly, increasing numbers of educators are tuning into the idea of play as something serious and rigorous. A Serious Play Conference is held annually by key game developers as well as educators; while Michigan State University offers a Masters of Arts in "Serious Games."

Play is not only not frivolous, but capable of producing serious intellectual work and an activity that possesses deep political power. Contrary to our commonplace understandings of play, I argue that a thoughtful analysis of the political power of play is potentially one of the most fruitful areas for those of us who are interested in furthering the "critical pedagogy" of Paulo Freire (2014)—a type of pedagogy that involves teaching both the oppressed and the oppressor of the structural mechanics that create these oppressions.

Why and how is play political? Let me illustrate with several examples. In Edward Said's introduction to the graphic novel *Palestine* by Joe Sacco (2003), an account of Sacco's experience of the daily struggles, humiliations, and frustrations of the Palestinians in the occupied territories, Said argues that "most adults ... tend to connect comics with what is frivolous or ephemeral, and there is an assumption that as one grows up, they are put aside for more serious pursuits" (p. i). He goes on to describe his first experience reading comic books—which were instantly banned by parents and school authorities—as one which made him feel at once "liberated and subversive." For the child Said, comics are a playful, seductive and dangerous agent:

> Everything about the enticing book of colored pictures, but specially its untidy, sprawling format, the colorful, riotous extravagance of its pictures, the unrestrained passage between what the characters thought and said, the exotic creatures and adventures reported and depicted: all this made up for a hugely wonderful thrill, entirely unlike anything I had hitherto known or experienced. (i)

Comics blur boundaries in their graphic, colorful excess, represented by the abundance of images and risqué flows in depictions of conversations.

Said goes on to reflect on the reasons behind the almost authoritarian ban on comics by his parents and by school authorities. He notes that the ostensible reason for the ban was that comics "interfered with one's schoolwork" (Sacco, 2003, p. ii) But for Said, the logic behind the ban was deeper and more psychologically subversive—comics in form and content provided the ability to *imagine, create, and live alternative realities*: imagining the alternative being the most political of acts. The liberationary aspect of comics lay in their ability to express the unexpressed, to give form to the formless:

> In ways that I still find fascinating to decode, comics in their relentless foregrounding—far more, say, than film cartoon or funnies, neither of which mattered much to me—seemed to say what couldn't otherwise be said, perhaps what wasn't permitted to be said or imagined, defying the ordinary processes of thought, which are policed, shaped and reshaped by all sorts of pedagogical as well as ideological pressures. I knew nothing of this then, but I felt that comics freed me to think and imagine and see differently. (p. ii)

In these words, Said gets to the core of how the play expressed within comics is so subversive; within their playfulness, their excess, they create a space for a world to be imagined differently—the key towards creating a space for this difference which can be one day translated into reality.

This idea—of play being at once both serious, and political—is by no means unique to Said. In "The Location of Brazil," Salman Rushdie (1991) declares: "*Play. Invent the World*" (p. 123; original emphasis). It is *through* play that the world is constructed and deconstructed—only to allow us the ability to imagine alternative forms of construction. Rushdie's essay is a deconstruction of the film *Brazil* by Terry Gilliam (1985), which depicts an Orwellian world controlled by meaningless bureaucracy and machines. The point of Rushdie's essay hinges on the question: What is the location of the Brazil that the film alludes to? There is no sign of the South American country in the film's grim dystopia outside of the film's haunting refrain: "Brazil… Where hearts were entertained in June / We stood beneath an amber moon / And softly whispered, someday soon."

The location of the Brazil in the film *Brazil* is key because it represents imagination, the creative impulse, and play; all three of which are deeply political as they allow the dreaming of alternate possibilities and realities. Recalling Joseph Conrad's injunction that the goal of the writer is to ultimately make one see differently outside of the usual *idées reçues*, Rushdie (1991) argues that "the true location of Brazil is the other great tradition in art, the one in which techniques of comedy, metaphor, heightened imagery, fantasy, and so on are used to break down our conventional, habit-dulled certainties about what the world is and has to be" (p. 122). In other words, just like the subversive power of play in the comic books that Said discovered, through breaking down our "conventional, habit-dulled certainties," *play reimagines the world*. To reimagine the world is to create the potential to change it.

Play is, in brief, serious business. To play is to imagine; to imagine is political because it allows us to envision a different order, a different system, a different way to separate economic resources and power. In this light, games are potentially extremely powerful because they go further in terms of forms of identification. Games function on a type of rhetorical power that differs from narrative, text, and image. Ian Bogost (2007) has termed this *procedural rhetoric*: "the art of persuasion through rule-based representations and interactions rather than the spoken word, writing, images or moving pictures" (p. ix). In his analysis, Bogost (2007) claims that the power of games for education does not stem from the *content* of games, but through "the very way videogames mount claims through procedural rhetorics" (p. ix). To put it simply: Games depend on a set of rules which then determine possible movements and outcomes. These movements and outcomes are *procedural*; and the act of performing these procedures has a *rhetorical* effect. To play a game, in this sense, involves being interpolated in an Althusserian sense into a type of subject position; the act of playing a game implicitly asks a player to accept a set of rules, and rhetorics which structure the world of the game, and movement through it.

25.1 Civilization: Colonization and The War on Terror

Games thus offer a different dimension to the declaration of Frantz Fanon (2008) that "to speak a language is to appropriate a world and culture" (p. 21). The languages of games are encoded in their rules, their dynamics; for Bogost and some others, these rules for games are further encapsulated by computational dynamics when it comes to video games. When you play a game, you not only take on the world and the culture it is based upon, you function by accepting the rules and possibilities open to you by the game. In this regard, seemingly politically neutral games are hardly such. The popular video game *Civilization: Colonization* by Sid Meier is an excellent example of this. The game, a riff of Sid Meier's original game *Civilization*, was first released in 1994 and updated in 2008. *Civilization: Colonization* puts players in the place of an European power and turns them loose on the Americas with a goal of colonizing the entire territory.

Trying to unpack some of the racial rhetorics of actually playing the game, Trevor Owens and Rebecca Mir (2012) modded the game to see what would happen if they tried to make indigenous characters playable. What they found out: The code that created the indigenous characters was written so differently from the European characters that they were basically useless and devoid of agency. Specifically, Owens and Mir (2012) note that "Natives were created by consciously turning off individual characteristics of standard peoples in *Civilization IV*. That is to say, Native peoples are not a different kind of entity in the game; they are quite literally another kind of people." In their analysis, Owens and Mir (2012) reveal that indigenous characters are rendered as disabled, agency-less characters, such that the "scripts speak them into existence at the level of code as a defunct, stripped and inhibited version of their oppressors."

To play *Civilization: Colonization*, thus is, in terms of content, to accept that the default human subject position is that of a European colonizer. Using the idea of procedural rhetoric, one can only play the game effectively if one plays as a colonizer. However, this level of analysis only covers the *content* portion of the game. When stripped down to the level of *code*, the procedural, computational rhetoric that determines the game shows that there is no way to "play" the Native characters that would even allow them to interact on the same level as the European characters; they have been deformed as objects as recreated within the ontology and epistemology of the game world.

Playing *Civilization: Colonization* unthinkingly would thus be to assume the world—the culture—of the colonizer as the default world. Understanding the function of this rhetoric of play is important—both in terms of appreciating the impact of the game, both at the level of content and code, as well as designing alternatives to it. Because games can be created not simply to reinforce colonial ideologies but also to subvert them. For example, *War on Terror, the Boardgame*, produced by TerrorBull Games in 2006, is a satire on the 2003 U.S. invasion of Iraq, inspired by the board game *Risk*. Through game mechanics, the game uses irony to show the effects of imperialism and geopolitics. Each player begins with a tiny presence on the world map as a budding empire, intent on "liberating" countries and continents, through controlling oil production and building cities. An empire manages to control a region when it has a development there (village, town, or city), and it can only build developments in a neighboring region if it is unoccupied. A player can interfere with others' attempts to build empire by either 1) fighting wars against them, or 2) funding terrorist units to attack their opponents. The game is card driven, where each card will allow you a variety of actions, and more cards (and actions) can be purchased through buying oil, which is randomly spread out on the map and differs each time the game is played. Importantly, once players are out of the game, they become "terrorists"—and can still influence the result and even win the game.

Much of the mechanics of the game also center around secret diplomacy and behind-the-scenes negotiations between different empires—the game includes a "secret message pad" to facilitate this. The game also includes an "Axis of Evil"—a spinner at the center of the board that determines which of the players is "evil" and has to wear the "Evil balaclava". All empires have a financial incentive to fight a war against the player that has randomly become designated as the "evil empire."

Through its game mechanics, then, *War on Terror: The Boardgame* deftly encourages its players to understand some of the hypocrisy and complex mechanisms behind the U.S. invasion of Iraq; that much of the rhetoric behind "liberation" through occupation is motivated by the financial incentives of controlling oil resources; that empire propaganda of "liberation" is often accompanied by covert funding of terrorism by empires themselves against their enemies; that the notion of an "evil empire" is a completely random one, but one which has deep political and financial consequences as other empires develop a financial incentive to invade the random player who has been designated as "evil." The game mechanics also teach the players that it is the dispossessed who are inclined to become terrorists, as they have nothing to lose—all players who have been dispossessed by other empires become terrorists in the end.

Games and play thus offer some of the most interesting and underutilized ways of creating political thought and action. Through their mechanics and rhetorics, they create worlds which ask players to internalize political ideologies on the levels of content, mechanic, design, code; by the ways actions in games are supported and delimited by their creators.

25.2 Trading Races and the Political Power of Play

Because of the tremendous power of games and play, I spent the 2012–13 academic year writing a role-playing game called *Trading Races*, which was supported by Stockton College and the Humanities Writ Large grant at Duke University. I wrote this game because I teach so many classes on gender, race, and ethnicity; on types of oppression which dominant groups find difficulty identifying with. I've been intrigued by the political power that I have seen embedded in games, in how they allow for different levels of identification and engagement, and wondered if I would be able to apply that to my classroom through playing a game.

Trading Races is set in 2003, right before the landmark decisions on affirmative action took place at the University of Michigan. Players in the game take on a combination of complex characters which range from real to imaginary, such as Sandra Day O'Connor, Clarence Thomas, and bell hooks, to members of a multi-ethnic, multi-national Michigan Student Assembly. Players are divided into three factions: color-blind (against affirmative action); color-conscious (for affirmative action); and indeterminate. Each character has a set of individual goals which remain secret from the rest of the other players, which consist of meeting their own individual objectives as well as faction objectives. There are players of the same ethnicity representing different parts of the political spectrum; for example, there are pro-affirmative action Asian American characters, anti-affirmative action Asian American characters, and undecided ones. The game takes about nine to ten class sessions, or about three to four weeks. There are three stages in the game—the pre-game period, where the instructor discusses the course material with the students, actual gameplay, where students debate in character, and the post-game analysis, where instructor and students reflect on the metadynamics behind the game and how it might have diverged from actual history.

Players win the game through effective rhetorical persuasion. The goal of the game is to convince others to agree with your character's position. For this reason, a person who plays this game well has to effectively demonstrate that she understands his or her character's ideological position so well that she or he can convince other players to side with her, whether she agrees with her given character's position personally or not. For this reason, this game is really about "trading ideological positions" rather than trading "races" per se. I derived my inspiration for this game from a series of educational role playing games first developed out of Barnard College ten years ago titled *Reacting to the Past*, all of which focus on games which take place at critical historical moments, such as the Partition of India, the French Revolution, the ending of Apartheid in South Africa. These games are called *Reacting* to the past, not *Re-enacting* the past, because in many cases, history unfolds differently in different iterations of the game when played; some important things that happened historically might not happen in these games, all of which are teaching mechanisms for the contingencies within social movements and histories.

Thus far I've been extremely impressed by the results of my game. *Trading Races* has been played several times—with two groups of undergraduates at Duke by my colleague Eileen Chow; at an upper-level history seminar with Sharon Musher, and by myself in Fall 2013 at Stockton College. When I taught the game as a class in the Fall, I spent the first half of the semester discussing the core readings with my students before gameplay actually began. My students blogged and produced an imaginary Michigan student newspaper during the game. I was impressed by how much their writing and speaking improved while roleplaying—it was markedly so much better than what I saw in the classroom when they were not in character. At the end of each of the playtests, various students from different classes have reported that while they have not fundamentally changed their ways of thinking, they have had to consider the nuances of the opposing side a lot more carefully than they initially did, which to me, speaks of pedagogical success.

I say again: Play is serious business. Play and games are immensely powerful in their ability to shape and create worlds through the building of platforms, rules and mechanics. This power is also political. Play and games have tremendous rhetorical power that can be harnessed for a plethora of unexplored social, political and pedagogical purposes. We need to deeply reconsider the role of play in how we educate ourselves and each other, because play is not merely the work of childhood as Mr. Rogers claims, but serious work that involves us all.

Ghost Towns of the Public Good

Pat Lockley

I never really got tenure as a concept, and after almost ten years of e-learning I finally found a job which didn't feature a ticking clock time bomb as its soundtrack. Sadly, in this sound of silence, came a new friend, a broken camel's back and I'd broken-become-beladen with an LMS-anthropy which was destined to push me away. Perhaps I have commitment issues, perhaps I'd spent so long searching for a brand that I'd grown tired of red-hot metal LinkedIn endorsements. I've never had so many valued skills, but found less demand for them. Part Bitcoin, part bit part. Be your own bubble.

Pop! I'd seen glorious projects I'd made die, in one instance the world's biggest open-education search engine—something you'd assume would be celebrated and honoured—slowly grind itself into obsoletion. You'd sneak out of the academy, flirt with organisations that felt so fresh, and find the same thing. You'd see stuff you built for free be replaced with a Kickstarter funded project because they had the funds, and they knew how to kick you in the teeth and when you're down. Often, even after leaving one institution, I'd find myself returning to fix things because I'd not been replaced. A sense of parental care drove me to maintain things I used to get paid to do. Time, though, made me a horrible father: Sometimes I would sit round and watch things die, or just leave and hope something would save them. Sometimes you had to let code fly the nest. Sometimes you took a stand. Sometimes you'd finish building before flirting with the next project.

So each contract was a sudden burst of joyous activity, and you could forget, maybe for a few beautiful months that it would die. Marrying the mayfly—doomed before it started—but you just thought if you believed it would work, that this time it could be different. But sooner or later someone would say three little words: *no money left*. All you could do was hope you'd made good enough memories. Maybe maybe maybe if you just worked harder, it would get through to higher management and they'd love it enough to give it more money. So work days stretched as hope faded, and you could rage—and how we would rage—against the coming of the night. Each failure, marked with more scar tissue (see the aforementioned spinal SnapPage), was another soul-sucking investigation into why it had happened again. You hoped those memories would be enough to counter how everything would become seen as a mistake, how we didn't make it sustainable, how there wasn't the demand, how they'd love to work together again, how they'd tried hard, how come it had to end.

The answer: Because it just does.

Maybe one day we'd get a contract for longer than a year, and then, we could show them. We could spend a year building something excellent. Then as programmers we'd have time to understand marketing, or promotion, or sustainability, or community development so we could make sure the project would have enough supporters that it'd not die.

And once or twice, we got close; we built a glorious bridge to allow people to make the land so often promised. The interface made sense, the tool was useful, it would actually work and achieve real benefits.

People would ignore the bridge

People would cross the water on stilts made of razor blades as you watched

People would swim over in lead costumes as you looked on

People would drink the sea just to walk across as you looked on

Every scar from every other memory made no sense as uglier, functionally poorer, cumbersome systems won out. You couldn't defeat the millions of a venture capitalist, you couldn't get round the manna-funded-marketing of eponymous foundations. Kickstarter finished things for you.

The reason: Because money just does.

And so like every fairy tale, the bridge has its use as the roof of a home for a troll, embittered by years of tiptoeing traduced by one giant stride, of waiting patiently for anyone to be interested in e-learning and then watching the MOOC sweep away all before it like low-cost locust conquistadors. Or always being the tortoise that never beats the hare. Big companies just can't listen, and they don't need to either. What good is a teacher's voice when lost in translation through a system which doesn't hear them. So you try to build something different, you try to placate the LMS hatred which seems to be everywhere. By when? Yesterday, or in under six months.

You could dream of someone "tweeting your project into a top ten things", one day you might even make a national paper. What good is forever if you can be replaced in an evening? Why stay when the first few months work best? Why not just drift...

Because our solution never taught millions, but you never asked it to

Because our solution didn't work on [wondergadgetnamegoeshere], because they are brand new

Because our solution doesn't work for you, but you never tell us why

Because you don't know anyone using us as if we all have "I <3 Microsoft" in our Twitter bios

Because the contract is six months as the institution is still uncertain

Because e-learning is old enough to draw a pension and it still remains a pilot or an experiment

Because outside of blogs and tweets we never ever really spoke to each other did we? We never even told each other what we wanted or what we could do. We ended up taking shots and stabs in the dark and expected no one to get hurt. *Tenure, temporary* are the words of those divided and conquered, and the cliches of university life. Conversations always turned to money, as if dollar signs replaced full stops, so much did funding dominate conversations. Even talking about an idea meant you'd need to know how to pay for it. You can run the bath for your eureka moment, but you'd better believe the water will be ice cold till you pay your way.

If you believe in a public good, or a public service, then how can that be married to a deadline, or funding? Innately we become volunteers, sporadically paid, occasionally valued, while the temples fill with traders.

How does a six month developer build for a future they'll never see. How does a year-long project not become another experiment—where experiment is just a justification for myopia, seeing everything through six-month eyes. Another gold rush, another ghost town, with ghost suburbs, ghost malls, and ghost schools. It doesn't take an expert to see the neglect inherent in short-term projects once the beans are counted. And we wonder why no one wants to use e-learning software? When you ask why it doesn't do X—the answer is no funding. Always no funding.

So why stick around to make such temporary things, when you can set yourself free to listen, to grow, to develop, to build sustainably? Everytime the money runs out, all that lost momentum starts to leave voids in need of filling. University work becomes more hole than thing. Unfinished sympathies. Nature abhors a vacuum, and silence is just a space noise hasn't found.

Ed-Tech in a Time of Trump

Audrey Watters

Funny word, "hope." Funny, those four letters used so iconically to describe a presidential campaign from a young Illinois senator, a campaign that seems now lifetimes ago. Hope.

My talks—and I guess I'll warn you in advance if you aren't familiar with my work—are not known for being full of hope. Or rather I've never believed the hype that we should put all our faith in, rest all our hope on technology. But I've never been hopeless. I've never believed humans are powerless. I've never believed we could not act or we could not do better.

There were a couple of days, following our decision about the title and topic of this keynote—"Ed-Tech in a Time of Trump"—when I wondered if we'd even see a Trump presidency. Would some revelation about his business dealings, his relationship with Russia, his disdain for the Constitution prevent his inauguration? We should have been so lucky, I suppose. Hope.

The thing is, I'd still be giving the much the same talk, just with a different title. "A Time of Trump" could be "A Time of Neoliberalism" or "A Time of Libertarianism" or "A Time of Algorithmic Discrimination" or "A Time of Economic Precarity." All of this is—from President Trump to the so-called "new economy"—has been fueled to some extent by digital technologies; and that fuel, despite what I think many who work in and around education technology have long believed—have long hoped—is not necessarily (heck, even remotely) progressive.

I've had a sinking feeling in my stomach about the future of education technology long before Americans—26% of them, at least—selected Donald Trump as our next President. I am, after all, "ed-tech's Cassandra." But President Trump has brought to the forefront many of the concerns I've tried to share about the politics and the practices of digital technologies. I want to state here at the outset of this talk: We should be thinking about these things no matter who is in the White House, no matter who runs the Department of Education (no matter even whether we have a federal department of education). We should be thinking about these things no matter who heads our university. We should be asking—always and again and again: Just what sort of future is this technological future of education that we are told we must embrace?

Of course, the *future* of education is always tied to its past, to the history of education. The future of technology is inexorably tied to its own history as well. This means that despite all the rhetoric about "disruption" and "innovation," what we find in technology is a layering onto older ideas and practices and models and systems. The networks of canals, for example, were built along rivers. Railroads followed the canals. The telegraph followed the railroad. The telephone, the telegraph. The Internet, the telephone and the television. The Internet is largely built upon a technological infrastructure first mapped and built for freight. It's no surprise the Internet views us as objects, as products, our personal data as a commodity.

When I use the word "technology," I draw from the work of physicist Ursula Franklin (1999) who spoke of technology as a practice: "Technology is not the sum of the artifacts, of the wheels and gears, of the rails and electronic transmitters," she wrote. "Technology is a *system*. It entails far more than its individual material components. Technology involves organization, procedures, symbols, new words, equations, and, most of all, a mindset" (pp. 2–3). "Technology also needs to be examined as an agent of power and control," Franklin insisted, and her work highlighted "how much modern technology drew from the prepared soil of the structures of traditional institutions, such as the church and the military" (p. 3).

I'm going to largely sidestep a discussion of the church today, although I think there's plenty we could say about faith and ritual and obeisance and technological evangelism. That's a topic for another keynote perhaps. And I won't dwell too much on the military either—how military industrial complexes point us towards technological industrial complexes (and to ed-tech industrial complexes in turn). But computing technologies undeniably carry with them the legacy of their military origins. Command. Control. Communication. Intelligence.

As Donna Haraway (1991) argues in her famous "Cyborg Manifesto," "Feminist cyborg stories have the task of recoding communication and intelligence to subvert command and control" (p. 175). I want those of us working in and with education technologies to ask if that is the task we've actually undertaken. Are our technologies or our stories about technologies feminist? If so, when? If so, how? Do our technologies or our stories work in the interest of justice and equity? Or, rather, have we adopted technologies for teaching and learning that are much more aligned with that military mission of command and control? The mission of the military. The mission of the church. The mission of the university.

I do think that some might hear Haraway's framing—a call to "recode communication and intelligence" (1991, p. 175)—and insist that that's exactly what education technologies do, and they do so in a progressive reshaping of traditional education institutions and practices. Education technologies facilitate communication, expanding learning networks beyond the classroom. And they boost intelligence—namely, how knowledge is created and shared.

Perhaps they do.

But do our ed-tech practices ever actually recode or subvert command and control? Do (or how do) our digital communication practices differ from those designed by the military? And most importantly, I'd say, does (or how does) our notion of intelligence?

"Intelligence"—this is the one to watch and listen for. (Yes, that's ironic that "ed-tech in a time of Trump" will be all about intelligence, but hear me out.)

"Intelligence" means understanding, intellectual, mental faculty. Testing intelligence, as Stephen Jay Gould (1996) and others have argued, has a long history of ranking and racism. The word "intelligence" is also used, of course, to describe the gathering and assessment of tactical information—information, often confidential information, with political or military value. The history of computing emerges from cryptography, tracking and cracking state secrets. And the word "intelligence" is now used—oh so casually—to describe so-called "thinking machines": algorithms, robots, AI.

It's probably obvious—particularly when we think of the latter—that our notions of "intelligence" are deeply intertwined with technologies. "Computers will make us smarter"—you know those assertions. But we've long used machines to measure and assess "intelligence" and to monitor and surveil for the sake of "intelligence." And again, let's recall Franklin's definition of technologies includes not just hardware or software, but ideas, practices, models, and systems.

One of the "hot new trends" in education technology is "learning analytics"—this idea that if you collect enough data about students that you can analyze it and in turn algorithmically direct students towards more efficient and productive behaviors, institutions towards more efficient and productive outcomes. Command. Control. Intelligence.

And I confess, it's that phrase "collect enough data about students" that has me gravely concerned about "ed-tech in a time of Trump." I'm concerned, in no small part, because students are often unaware of the amount of data that schools and the software companies they contract with know about them. I'm concerned because students are compelled to use software in educational settings. You can't opt out of the learning management system. You can't opt out of the student information system. You can't opt out of required digital textbooks or digital assignments or digital assessments. You can't opt out of the billing system or the financial aid system. You can't opt of having your cafeteria purchases, Internet usage, dorm room access, fitness center habits tracked. Your data as a student is scattered across multiple applications and multiple databases, most of which I'd wager are not owned or managed by the school itself but rather outsourced to a third-party provider.

School software (and I'm including K–12 software here alongside higher ed) knows your name, your birth date, your mailing address, your home address, your race or ethnicity, your gender (I should note here that many education technologies still require "male" or "female" and do not allow for alternate gender expressions). It knows your marital status. It knows your student identification number (it might know your Social Security Number). It has a photo of you, so it knows your face. It knows the town and state in which you were born. Your immigration status. Your first language and whether or not that first language is English. It knows your parents' language at home. It knows your income status—that is, at the K–12 level, if you quality for a free or reduced lunch and at the higher ed level, if you qualify for a Pell Grant. It knows if you are the member of a military family. It knows if you have any special education needs. It knows if you were identified as "gifted and talented." It knows if you graduated high school or passed a high school equivalency exam. It knows your attendance history—how often you miss class as well as which schools you've previously attended. It knows your behavioral history. It knows your criminal history. It knows your participation in sports or other extracurricular activities. It knows your grade level. It knows your major. It knows the courses you've taken and the grades you've earned. It knows your standardized test scores.

Obviously it's not a new practice to track much of that data, and as such these practices are not dependent entirely on new technologies. There are various legal and policy mandates that have demanded for some time now that schools collect this information. Now we put it in "the cloud" rather than in a manila folder in a locked file cabinet. Now we outsource this to software vendors, many of whom promise that because of the era of "big data" that we should collect even more information about students—all their clicks and their time spent "on task," perhaps even their biometric data and their location in real time—so as to glean more and better insights. Insights that the vendors will then sell back to the school.

Big data.

Command. Control. Intelligence.

This is the part of the talk, I reckon, when someone who speaks about the dangers and drawbacks of "big data" turns the focus to information security and privacy. No doubt schools are incredibly vulnerable on the former front. Since 2005, U.S. universities have been the victim of over 1,300 data breaches involving more than 24 million known records (Cook, 2020). We typically think of these hacks as going after Social Security Numbers or credit card information or something that's of value on the black market.

The risk isn't only hacking. It's also the rather thoughtless practices of information collection, information sharing, and information storage. Many software companies claim that the data that's in their systems is *their* data. It's questionable if much of this data—particularly metadata—is covered by FERPA. As such, student data can be sold and shared, particularly when the contracts signed with a school do not prevent a software company from doing so. Moreover, these contracts often do not specify how long student data can be kept.

In this current political climate—ed-tech in a time of Trump—I think universities need to realize that there's a lot more at stake than just financially motivated cybercrime. Think Wikileaks' role in the presidential election, for example. Now think about what would happen if the contents of your email account was released to the public. President Trump has made it a little bit easier, perhaps, to come up with "worse case scenarios" when it comes to politically-targeted hacks, and we might be able to imagine these in light of all the data that higher-ed institutions have about students (and faculty).

Again, the risk isn't only hacking. It's amassing data in the first place. It's profiling. It's tracking. It's surveilling. It's identifying "students at risk" and students who are "risks."

Back when I was first working as a freelance tech journalist, I interviewed an author about a book he'd written on big data and privacy. He made one of those casual remarks that you hear quite often from people who work in computing technologies: Privacy is dead. He'd given up on the idea that privacy was possible or perhaps even desirable; what he wanted instead was transparency—that is, to know who has your data, what data, what they do with it, who they share it with, how long they keep it, and so on. You can't really protect your data from being "out there," he argued, but you should be able to keep an eye on where "out there" it exists.

This particular author reminded me that we've counted and tracked and profiled people for decades and decades and decades and decades. In some ways, that's the project of the census—first conducted in the United States in 1790. It's certainly the project of much of the data collection that happens at school. And we've undertaken these practices since well before there was "big data" or computers to collect and crunch it. Then he made a comment that, even at the time, I found deeply upsetting. "Just as long as we don't see a return of Nazism," he joked, "we'll be okay. Because it's pretty easy to know if you're a Jew. You don't have to tell Facebook. Facebook knows."

We can substitute other identities there. It's easy to know if you're Muslim. It's easy to know if you're queer. It's easy to know if you're pregnant. It's easy to know if you're Black or Latino or if your parents are Syrian or French. It's easy to know your political affinities. And you needn't have given over that data, you needn't have "checked those boxes" in your student information system in order for the software to develop a fairly sophisticated profile about you.

Think of a punch card, a paper-based method of proto-programming, one of the earliest ways in which machines could be automated. It's a relic, a piece of "old tech," if you will, but it's also a political symbol. Think draft cards. Think the slogan "Do not fold, spindle or mutilate." Think Mario Savio on the steps of Sproul Hall at UC Berkeley in 1964, insisting angrily that students not be viewed as raw materials in the university machine.

The first punch cards were developed to control the loom, industrializing the craft of weaving women around 1725. The earliest design—a paper tape with holes punched in it—was improved upon until the turn of the 19th century, when Joseph Marie Jacquard first demonstrated a mechanism to automate loom operation.

Jacquard's invention inspired Charles Babbage, often credited with originating the idea of a programmable computer. A mathematician, Babbage (1982, p. 26) believed that "number cards," "pierced with certain holes," could operate the Analytical Engine, his plans for a computational device. "We may say most aptly that the Analytical Engine weaves algebraical patterns just as the Jacquard-loom weaves flowers and leaves," Ada Lovelace, Babbage's translator and the first computer programmer, wrote (Menebrea, 1837, p. 696).

But it was Herman Hollerith who invented the recording of data on this medium so that it could then be read by a machine. Earlier punch cards—like those designed by Jacquard—were used to control the machine. They weren't used to store data. But Hollerith did just that. The first Hollerith card had 12 rows and 9 columns, and data was recorded by the presence or absence of a hole at a specific location on a card.

Hollerith founded The Tabulating Machine Company in 1896, one of four companies consolidated to form Computing-Tabulating-Recording Company, later renamed the International Business Machines Corporation, or IBM.

Hollerith's punch card technology was first used in the U.S. Census in 1890 to record individual's traits—their gender, race, nationality, occupation, age, marital status. These cards could then be efficiently sorted to quantify the nation. The census was thrilled as it had taken almost a decade to tabulate the results of the 1880 census, and by using the new technology, the agency saved $5 million.

Hollerith's machines were also used by Nicholas II, the czar of Russia for the first (and only) census of the Russian Imperial Empire in 1897. And they were adopted by Hitler's regime in Germany. As Edwin Black (2001) chronicles in his book *IBM and the Holocaust*:

> When Hitler came to power, a central Nazi goal was to identify and destroy Germany's 600,000-member Jewish community. To Nazis, Jews were not just those who practiced Judaism, but those of Jewish blood, regardless of their assimilation, intermarriage, religious activity, or even conversion to Christianity. Only after Jews were identified could they be targeted for asset confiscation, ghettoization, deportation, and ultimately extermination. To search generations of communal, church, and governmental records all across Germany—and later throughout Europe—was a cross-indexing task so monumental, it called for a computer. But in 1933, no computer existed. (p. 8)

What did exist at the time was the punch card and the IBM machine, sold to the Nazi government by the company's German subsidiary, Dehomag.

Hitler's regime made it clear from the outset that it was not interested in merely identifying those Jews who claimed religious affiliation, who said that they were Jewish. It wanted to be able to find those who had Jewish ancestry, Jewish "blood," those who were not Aryan.

Hitler called for a census in 1933, and Germans filled out the census on pen and paper—one form per household. There was a census again in 1939, and as the Third Reich expanded, so did the Nazi compulsion for data collection. Census forms were coded and punched by hand and then sorted and counted by machine—*IBM* punch cards and *IBM* machines. During its relationship with the Nazi regime—one lasting throughout Hitler's rule, throughout World War II—IBM derived about a third of its profits from selling punch cards.

Column 22 on the punch card was for religion—punched at hole 1 to indicate Protestant, hole 2 for Catholic, hole 3 for Jew. The Jewish cards were processed separately. The cards were sorted and indexed and filtered by profession, national origin, address, and other traits. The information was correlated with other data—community lists, land registers, medical information—in order to create a database, "a profession-by-profession, city-by-city, and indeed a block-by-block revelation of the Jewish presence" (Black, 2001, p. 58).

It was a database of inference, relying heavily on statistics alongside those IBM machines. This wasn't just about those who'd "ticked the box" that they were Jewish. Nazi "race science" believed it could identify Jews by collecting and analyzing as much data as possible about the population. "The solution is that every interesting feature of a statistical nature … can be summarized … by one basic factor," the Reich Statistical Office boasted. "This basic factor is the Hollerith punch card" (qtd. in Black, 2001, p. 94).

Command. Control. Intelligence.

The punch card and the mechanized processing of its data were used to identify Jews, as well as Roma and other "undesirables" so they could be imprisoned, so their businesses and homes could be confiscated, so their possessions could be inventoried and sold. The punch card and the mechanized processing of its data was used to determine which "undesirables" should be sterilized, to track the shipment of prisoners to the death camps, and to keep tabs on those imprisoned and sentenced to die therein. All of this recorded on IBM punch cards and IBM machines.

The CEO of IBM at this time, by the way: Thomas Watson. Yes, this is who IBM has named their "artificial intelligence" product Watson after—IBM Watson, which has partnered with Pearson and with Sesame Street, to "personalize learning" through data collection and data analytics.

Now a quick aside, since I've mentioned Nazis.

Back in 1990, in the early days of the commercialized Internet, those heady days of Usenet newsgroup discussion boards, attorney Mike Godwin (1994) "set out on a project in memetic engineering." Godwin felt as though comparisons to Nazis occurred too frequently in online discussions. He believed that accusations that someone or some idea was "Hitler-like" were thrown about too carelessly. "Godwin's Law," as it came to be known, says that "As an online discussion grows longer, the probability of a comparison involving Hitler approaches 1" (Godwin, 1994). Godwin's Law has since been invoked to decree that once someone mentions Hitler or Nazis, that person has lost the debate altogether. Pointing out Nazism online is off-limits.

Perhaps we can start to see now how dangerous, how damaging to critical discourse this even rather casual edict has been.

Let us remember the words of Supreme Court Justice Robert Jackson in his opening statement for the prosecution at the Nuremburg Trials (International Military Tribunal, 1947):

> What makes this inquest significant is that these prisoners represent sinister influences that will lurk in the world long after their bodies have returned to dust. We will show them to be living symbols of racial hatreds, of terrorism and violence, and of the arrogance and cruelty of power. ... Civilization can afford no compromise with the social forces which would gain renewed strength if we deal ambiguously or indecisively with the men in whom those forces now precariously survive.

We need to identify and we need to confront the ideas and the practices that are the lingering legacies of Nazism and fascism. We need to identify and we need to confront them in our technologies. Yes, in our education technologies. Remember: our technologies *are* ideas; they *are* practices. Now is the time for an ed-tech antifa, and I cannot believe I have to say that out loud to you.

And so you hear a lot of folks in recent months say "read Hannah Arendt." And I don't disagree. Read Arendt. Read *The Origins of Totalitarianism* (Arendt, 1973). Read her reporting from the Nuremberg Trials.

But also read James Baldwin. Also realize that this politics and practice of surveillance and genocide isn't just something we can pin on Nazi Germany. It's actually deeply embedded in the American experience. It is part of this country *as a technology*.

Let's think about that first U.S. census, back in 1790, when federal marshals asked for the name of each head of household as well as the numbers of household members who were free white males over age 16, free white males under 16, free white females, other free persons, and slaves. In 1820, the categories were free white males, free white female, free colored males and females, and slaves. In 1850, the categories were white, Black, Mulatto, Black slaves, Mulatto slaves. In 1860, white, Black, Mulatto, Black slaves, Mulatto slaves, Indian. In 1870, white, Black, Mulatto, Indian, Chinese. In 1890, white, Black, Mulatto, Quadroon, Octoroon, Indian, Chinese,

Japanese. In 1930, white, Negro, Indian, Chinese, Japanese, Filipino, Korean, Hindu, Mexican.

You might see in these changing categories a changing demographic; or you might see this as the construction and institutionalization of categories of race—particularly race set apart from a whiteness of unspecified national origin, particularly race that the governing ideology and governing system wants identified and wants managed. The construction of Blackness. "Census enumeration is a means through which a state manages its residents by way of formalized categories that fix individuals within a certain time and a particular space," as Simone Browne (2015) writes in her book *Dark Matters: On the Surveillance of Blackness*, "making the census a technology that renders a population legible in racializing as well as gendering ways" (p. 56). It is "a technology of disciplinary power that classifies, examines, and quantifies populations" (p. 57).

Command. Control. Intelligence.

Does the data collection and data analysis undertaken by schools work in a similar way? How does the data collection and data analysis undertaken by schools work? What bodies and beliefs are constituted therein? Is whiteness and maleness always there as "the norm" against which all others are compared? Are we then constructing and even naturalizing certain bodies and certain minds as "undesirable" bodies and "undesirable" minds in the classroom, in our institutions by our obsession with data, by our obsession with counting, tracking, and profiling?

Who are the "undesirables" of ed-tech software and education institutions? Those students who are identified as "cheats," perhaps. When we turn the cameras on, for example with proctoring software, those students whose faces and gestures are viewed—visually, biometrically, algorithmically—as "suspicious." Those students who are identified as "out of place." Not in the right major. Not in the right class. Not in the right school. Not in the right country. Those students who are identified—through surveillance and through algorithms—as "at risk." At risk of failure. At risk of dropping out. At risk of not repaying their student loans. At risk of becoming "radicalized." At risk of radicalizing others. What about those educators at risk of radicalizing others. Let's be honest with ourselves, ed-tech in a time of Trump will undermine educators as well as students; it will undermine academic freedom. It's already happening. See Trump's tweets about Berkeley on February 2, 2017.

What do schools do with the capabilities of ed-tech as surveillance technology now in the time of a Trump? The proctoring software and learning analytics software and "student success" platforms all market themselves to schools claiming that they can truly "see" what students are up to, that they can predict what students will become. ("How will this student affect our averages?") These technologies claim they can identify a "problem" student, and the implication, I think, is that then someone at the institution "fixes" her or him. Helps the student graduate. Convinces the student to leave.

But these technologies do not *see* students. And sadly, *we* do not see students. This is cultural. This is institutional. We do not see who is struggling. And let's ask why we think, as the New York Times argued today, we need big data to make sure students graduate. Universities have not developed or maintained practices of compassion. Practices are technologies; technologies are practices. We've chosen computers instead of care. (When I say "we" here I mean institutions not individuals within institutions. But I mean some individuals too.) Education has chosen "command, control, intelligence." Education gathers data about students. It quantifies students. It has adopted a racialized and gendered surveillance system—one that committed to disciplining minds and bodies—through our education technologies, through our education practices.

All along the way, or perhaps somewhere along the way, we have confused surveillance for care.

And that's my takeaway for folks here today: When you work for a company or an institution that collects or trades data, you're making it easy to surveil people and the stakes are high. They're always high for the most vulnerable. By collecting so much data, you're making it easy to discipline people. You're making it easy to control people. You're putting people at risk. You're putting students at risk.

You can delete the data. You can limit its collection. You can restrict who sees it. You can inform students. You can encourage students to resist. Students have always resisted school surveillance.

But I hope that you also think about the culture of school. What sort of institutions will we have in a time of Trump? Ones that value open inquiry and academic freedom? I swear to you this: More data will not protect you. Not in this world of "alternative facts," to be sure. Our relationships to one another, however, just might. We must rebuild institutions that value humans' minds and lives and integrity and safety. And that means, in its current incarnation at least, in this current climate, ed-tech has very very little to offer us.

4 Reference

About the Authors 177
Permissions and Attributions 181
Bibliography 185
Index 203

About the Authors

Timothy R. Amidon is an associate professor at Colorado State University. He studies writing, professional and technical communication, and digital rhetoric, particularly intellectual property, multimodal literacies, and communication design.

Maha Bali / مها بالي is Associate Professor of Practice, Center for Learning and Teaching, American University in Cairo. She is co-founder of virtuallyconnecting.org and a facilitator of edcontexts.org.

Stephen R. Barnard is Associate Professor and Chair of the Sociology Department at St. Lawrence University in Canton, NY. He studies the role media and communication technologies play in relations of power, practice, and professionalism, and their implications for journalism and democracy.

Lee Skallerup Bessette is a learning design specialist in the Center for New Designs in Learning and Scholarship at Georgetown University. You can read more of her work at readywriting.org.

Marisol Brito is Assistant Professor of Philosophy at Metropolitan State University. Marisol's work focuses on race, gender, education and generally rethinking the world—especially education at all ages. Being a parent has made her a better learner, teacher and human.

Martha Fay Burtis is a Learning and Teaching Developer at Plymouth State University. She has worked in and around the intersection of higher ed, open pedagogy, and digital technologies for 20+ years.

Ian Derk is a contingent instructor of communication at a large university in the American Southwest. Despite hating hot weather, Ian has spent most of his life in deserts.

Robin DeRosa is Professor of English and Chair of Interdisciplinary Studies at Plymouth State University. She researches and writes about public university missions, OER, and open pedagogy.

Alex Fink is a Researcher and Learning Partner in the Youth Studies and Youth Development Leadership program at the University of Minnesota where he supports young people to lead change toward a socially just world.

Joseph P. Fisher is Executive Director of the Academic Resource Center at Georgetown University. Joe is a seasoned professor, with 16 years' experience at the community college and four-year university levels.

Chris Friend is an Assistant Professor of English, Director of *Hybrid Pedagogy*, and producer/host of the *HybridPod* podcast.

Chris Gilliard is a writer, professor and speaker. His scholarship concentrates on digital privacy, surveillance, and the intersections of race, class, and technology. He is an advocate for critical and equity-focused approaches to tech in education.

Abby Goode is Assistant Professor of English at Plymouth State University. She teaches courses on American literature, environmental writing, food studies, and critical theory.

Amy A. Hasinoff is Associate Professor of Communication at the University of Colorado Denver. She studies new media, gender, and sexuality.

Asao B. Inoue is a professor and the associate dean for Academic Affairs, Equity, and Inclusion in the College of Integrative Sciences and Arts at Arizona State University. His research focuses on antiracist and social justice theory and practices in writing assessments.

Adeline Koh is a former academic. She quit her tenured position as English professor to found Sabbatical Beauty, an artisanal, ethically manufactured Asian skincare brand, in 2017.

Tiffany Kraft is training organizer supporting the development and delivery of free, high-quality education for caregivers in Oregon through a trauma-informed lens.

Amanda Licastro is the Emerging & Digital Literacy Instructional Designer at The University of Pennsylvania Libraries. Her research explores the intersection of technology and writing, including book history, dystopian literature, and digital humanities, with a focus on multimodal composition and extended reality.

Ioana Literat is an Assistant Professor in the Communication, Media & Learning Technologies Design program at Teachers College, Columbia University. Her research examines creative participation in online contexts, with a particular focus on youth.

Pat Lockley is peripatetic. Interests: sailing closer to the wind, walking closer to the truth. I make things with love and hope, believing a better world is possible. Time to trespass. Impersonates a penguin online. Search for *pgogy*.

Maggie (Marijel) Melo is Assistant Professor in the School of Information and Library Science at UNC Chapel Hill and Director of Equity in the Making Lab.

Luca Morini is Research Fellow at Coventry University's Centre for Global Learning. He studies critical pedagogy, global perspectives on education, decolonisation, systems thinking, ecology, playfulness, education technology and co-creation. It all makes sense as a whole, he swears. :)

Sean Michael Morris is the Director of Digital Pedagogy Lab and Senior Instructor in Learning, Design, and Technology at the University of Colorado Denver.

Leif Nelson is the Director of Learning Technology Solutions at Boise State University. He studies the history of education and educational technology and the economic, social, and political tensions they (re)present.

Sherri Spelic is an educator, leadership coach and digital interloper at home in Vienna, Austria. She is the founder and publishing editor of Identity, Education, and Power.

Jesse Stommel is co-founder of Digital Pedagogy Lab and Hybrid Pedagogy. He is a documentary filmmaker and teaches courses about pedagogy, film, and new media.

Audrey Watters is an education writer, recovering academic, serial dropout, and part-time badass.

Jessica Zeller is an Associate Professor of Dance in the School for Classical & Contemporary Dance at Texas Christian University.

Permissions and Attributions

Most of the peceeding chapters are previously published work, reprinted with permission of their respective authors. Unless otherwise noted, material was initially published on *Hybrid Pedagogy*.

Foreword by Robin DeRosa; photo by Marina Shatskih on Unsplash
Introduction by Chris Friend; photo by John Peters on Unsplash

1. "Technology is Not Pedagogy" by Sean Michael Morris
   - Text originally appeared on the author's blog 10 June 2020
   - Photo by James Resly on Unsplash
2. "Building Castles in the Air" by Stephen Barnard
   - Text originally published 01 Sep 2015
   - "Pocket knife on edge" by Steve Snodgrass on Flickr; ⊜①
3. "Pedagogy, Prophecy, Disruption" by Ian Derk
   - Text originally published 17 Sep 2014
   - "itself" by Fio on Flickr; ⊜①⊚
4. "Slow Interdisciplinarity" by Abby Goode
   - Text originally published 12 Nov 2019
   - "Thick rope knot" by Robert Zunikoff on Unsplash
5. "The Process of Becoming" by Marisol Brito and Alexander Fink
   - Text originally published 02 Jul 2013
   - "Crayons" by Rafael J M Souza on Flickr; ⊜①
6. "Learning to Let Go" by Chris Friend
   - Text originally published 10 Sep 2014
   - "human" by Fio on Flickr; ⊜①⊚
7. "Seeking Patterns, Making Meaning" by Sherri Spelic
   - Text originally published 30 May 2017
   - "Castleton Mill" © Mark Stevenson on Flickr; reprinted with express permission of the photographer
8. "Messy and Chaotic Learning" by Martha Burtis
   - Text presented 31 Mar 2017 at Keene State College; published 05 Apr 2017 on *The Fish Wrapper*; abridged for print
   - "What does the future hold?" by Charles Deluvio on Unsplash
9. "Pedagogical Violence and Language Dominance" by Maggie Melo
   - Text originally published 03 May 2018
   - "Tuscan Lily" by Michael Taggart on Flickr; ⊜①⊚
10. "Trust, Agency, and Learning" by Jesse Stommel
    - Text originally published 16 Oct 2014 on HASTAC; expanded 09 Nov 2014 on *Hybrid Pedagogy*; revised for print
    - "view of the city" by Fio on Flickr; ⊜①⊚

11 "Confessions of a Subversive Student" by Leif Nelson
- Text originally published 04 Feb 2014
- "The world is wrong side up." by Eric May on Flickr; ⓒⓘⓢ

14 "A Soliloquy on Contingency" by Joseph P. Fisher
- Text originally published 15 Apr 2014
- "mic" by Robert Bejil on Flikr; ⓒⓘ

15 "N=1: Inquiry into Happiness and Academic Labor" by Ioana Literat
- Text originally published 04 Nov 2013
- "Nowhere" by Daniel Jensen on Unsplash

16 "From Ph.D. to Poverty" by Tiffany Kraft
- Text originally published 02 Sep 2014
- "let's sleep again" by Fio on Flickr; ⓒⓘⓢ

13 "On Silence" by Audrey Watters
- Text originally appeared on *Hack Education* 16 Aug 2014; reprinted by permission
- "Speak It Out Loud!" by Tobias Senkbeil on Flickr; ⓒⓘⓢⓞ

12 "Do You Trust Your Students?" by Amy Hasinoff
- Text originally published 22 Aug 2018
- "Instant" by Karina Vorozheeva on Unsplash

17 "When One Class is Not Enough" by Amanda Licastro
- Text original to this publication
- Colorful Hands 2 of 3 by Tim Mossholder on Unsplash

18 "Assessing so That People Stop Killing Each Other" by Asao B. Inoue
- Text originally appeared in Inoue, Asao B. (2019). *Labor-Based Grading Contracts: Building Equity and Inclusion in the Compassionate Writing Classroom.* Perspectives on Writing. The WAC Clearinghouse; University Press of Colorado. doi:10.37514/PER-B.2019.0216
- Photo by Manuela Böhm on Unsplash

19 "Pedagogy as Protest" by Jessica Zeller
- Text originally appeared on Jessica Zeller's blog 14 Jun 2020; reprinted with author's permission.
- Photo by Philippe Leone on Unsplash

25 "The Political Power of Play" by Adeline Koh
- Text originally published 03 Apr 2014
- "hopscotch" by nandadevieast on Flickr; ⓒⓘⓢ

21 "Interrogating the Digital Divide" by Lee Skallerup Bessette
- Text originally published 02 Jul 2012
- "To Educate" by Aaron Knox on Flickr; ⓒⓘⓢⓞ

24 "Education as Bulwark of Uselessness" by Luca Morini
- Text originally published 19 Jul 2016
- "Playful" by Tom Mrazek on Flickr; ☉①

20 "Critical Citizenship for Critical Times" by Maha Bali / مها بالي
- Text originally appeared on Al-Fanar Media 19 Aug 2013; reprinted with publisher and author permission
- Photo by Baher Khairy on Unsplash

22 "Pedagogy and the Logic of Platforms" by Chris Gilliard
- Text originally published in *Educause Review* 03 July 2017, ☉①☉; reprinted with author permission
- "The City Beneath Your Feet" by Jachan DeVol on Unsplash

23 "(dis)Owning Tech" by Timothy R. Amidon
- Text originally published 08 Sep 2016
- "ghost trestle" by Jason Carpenter on Flickr; ☉①☉☉

26 "Ghost Towns of the Public Good" by Pat Lockley
- Text originally published 09 Dec 2013
- "Crushed" by KayVee.INC on Flickr; ☉①☉☉

27 "Ed-Tech in a Time of Trump" by Audrey Watters
- Text originally appeared on *Hack Education* 02 Feb 2017; reprinted by permission
- Image courtesy Algorotoscope

Bibliography

Ad Fontes Media. (2021, January). *Interactive media bias chart.* https://www.adfontesmedia.com/interactive-media-bias-chart-2/. (Ctd. on 101)

Adichie, C. N. (2013). *Americanah.* Doubleday. (Ctd. on 102, 103).

Allen, R. L., & Rossatto, C. A. (2009). Does critical pedagogy work with privileged students? *Teacher Education Quarterly, 36*(1), 163–180 (Ctd. on 56).

Allen, W. (1975). *God.* S. French. (Ctd. on 70).

American Psychological Association. (2019). The decline of empathy and the rise of narcissism [podcast episode]. *Speaking of Psychology,* (95). https://www.apa.org/research/action/speaking-of-psychology/empathy-narcissism (Ctd. on 100)

Andersen, K. (2020). *Evil geniuses: The unmaking of america: A recent history.* Random House. (Ctd. on xiii, xvii).

Anderson, W. (Director). (1998). *Rushmore* [Film]. Touchstone. (Ctd. on 70).

Anthropy, A. (2012). *Rise of the videogame zinesters: How freaks, normals, amateurs, artists, dreamers, drop-outs, queers, housewives, and people like you are taking back an art form.* Seven Stories Press. (Ctd. on 147).

Aoun, J. E. (2017). *Robot-proof: Higher education in the age of artificial intelligence.* MIT press. (Ctd. on 19, 20).

APM Research Lab Staff. (2021, January 7). *The color of coronavirus: COVID-19 deaths by race and ethnicity in the U.S.* APM Research Lab. Retrieved January 17, 2021, from https://www.apmresearchlab.org/covid/deaths-by-race. (Ctd. on vii)

Arendt, H. (1973). *The origins of totalitarianism.* Houghton Mifflin Harcourt. (Ctd. on 171).

Arendt, H. (2013). *The human condition.* University of Chicago Press. (Ctd. on 110).

Arnab, S., Morini, L., & Clarke, S. (2018). Co-creativity with playful and gameful inspirations. *Proceedings of 12th European Conference on Game-Based Learning,* 1–7 (Ctd. on 144).

Arora, G., & Milk, C. (2015). *Clouds over Sidra* [Video]. Within. https://www.with.in/watch/CKRc5WA. (Ctd. on 103)

Assad, R., & Barsoum, G. (2007). *Youth exclusion in Egypt: In search of 'second chances'* (Middle East Youth Initiative Working Paper Number 2). Wolfensohn Center for Development, Dubai School of Government. (Ctd. on 124).

Babbage, C. (1982). On the mathematical powers of the calculating engine. In B. Randell (Editor), *The origins of digital computers: Selected papers* (3rd edition, Pages 19–54). Springer Berlin Heidelberg. https://doi.org/10.1007/978-3-642-61812-3_2. (Ctd. on 168)

Bali, M. (2013). *Critical thinking in context: Practice at an American liberal arts university in Egypt* (Doctoral dissertation). University of Sheffield. http://etheses.whiterose.ac.uk/4646/. (Ctd. on 121, 124)

Bali, M. (2015a). Embracing subjectivity. *Hybrid Pedagogy*. http://hybridpedagogy.org/embracing-subjectivity (Ctd. on 60)

Bali, M. (2015b). Pedagogy of care—gone massive. *Hybrid Pedagogy*. https://hybridpedagogy.org/pedagogy-of-care-gone-massive/ (Ctd. on 61)

Bali, M. (2015c). Yearning for praxis: Writing and teaching our way out of oppression. *Hybrid Pedagogy*. https://hybridpedagogy.org/yearning-for-praxis (Ctd. on 58)

Barnard, S. R. (2013). The case for technorealism [blog post]. *Stephen R. Barnard*. https://www.stephenrbarnard.com/blog/the-case-for-technorealism (Ctd. on 8)

Barnard, S. R., & Van Gerven, J. P. (2009). A people's method(ology): A dialogical approach. *Cultural Studies ↔ Critical Methodologies*, 9(6), 816–831 (Ctd. on 9).

Bateson, G. (1979). *Mind and nature: A necessary unity*. Dutton. (Ctd. on 145, 147).

Bayers, L., & Camfield, E. (2018). Student shaming and the need for academic empathy. *Hybrid Pedagogy* (Ctd. on 76).

Beck, E. (2014). Breaking up with Facebook: Untethering from the ideological freight of online surveillance. *Hybrid Pedagogy*. https://hybridpedagogy.org/breaking-facebook-untethering-ideological-freight-online-surveillance/ (Ctd. on 17)

Belenky, M. F., Clinchy, B. M., Goldberger, N. R., & Tarule, J. M. (1986). *Women's ways of knowing: The development of self, voice, and mind*. Basic Books. (Ctd. on 123).

Berg, M., & Seeber, B. K. (2016). *The slow professor: Challenging the culture of speed in the academy*. University of Toronto Press. (Ctd. on 20).

Bérubé, M. (2014). One last post about the MLA [blog post]. *Where I can post things with "hyper" links*. Retrieved February 1, 2021, from http://forathingthatdoesntfitonfacebook.wordpress.com (Ctd. on 86)

Bessette, L. S. (2013). On contingency, vocation, and loyalty [blog post]. *College Ready Writing*. https://www.insidehighered.com/blogs/college-ready-writing/contingency-vocation-and-loyalty (Ctd. on 87)

Black, E. (2001). *IBM and the Holocaust: The strategic alliance between Nazi Germany and America's most powerful corporation*. Random House. https://www.nytimes.com/books/first/b/black-ibm.html. (Ctd. on 169, 170)

Blankenship, L. (2019). *Changing the subject: A theory of rhetorical empathy*. University Press of Colorado. (Ctd. on 100, 104, 105).

Bloom, P. (2017). *Against empathy: The case for rational compassion*. Random House. (Ctd. on 105).

Bogost, I. (2007). *Persuasive games: The expressive power of videogames*. MIT Press. (Ctd. on 154).

Bogost, I. (2010). The turtlenecked hairshirt: Fetid and fragrant futures for the humanities. *Ian Bogost*. http://bogost.com/writing/blog/the_turtlenecked_hairshirt/ (Ctd. on 14)

Boice, B. (1996). Classroom incivilities. *Research in higher education, 37*(4), 453–486 (Ctd. on 77).

Bok, D. (2004). *Universities in the marketplace: The commercialization of higher education* (Volume 49). Princeton University Press. (Ctd. on 135).

Brenkert, G. G. (1981). Marx's critique of utilitarianism. *Canadian Journal of Philosophy, 11*(1), 193–220 (Ctd. on 146).

Brown, B. (2015). *Daring greatly: How the courage to be vulnerable transforms the way we live, love, parent, and lead*. Penguin. (Ctd. on 33).

Brown, S. (2017). Where every student is a potential data point. *The Chronicle of Higher Education* (Ctd. on 130).

Browne, S. (2015). *Dark matters: On the surveillance of Blackness*. Duke University Press. (Ctd. on 172).

Butler, R., & Nisan, M. (1986). Effects of no feedback, task-related comments, and grades on intrinsic motivation and performance. *Journal of Educational Psychology, 78*(3), 210 (Ctd. on 76).

Caillois, R. (2001). *Man, play, and games*. University of Illinois press. (Ctd. on 148).

Caulfield, M. (2017). Can higher education save the Web? EDUCAUSE *Review* (Ctd. on 129).

CCCC-IP Caucus. (2006). *CCCC-IP Caucus recommendations regarding academic integrity and the use of plagiarism detection services* [position statement]. Retrieved%20from%20http://culturecat.net/files/CCCC-IPpositionstatementDraft.pdf. (Ctd. on 139)

Chavez, N., & McKirdy, E. (2017). Look! Up in the sky! It's a new kind of cloud. *CNN Travel* (Ctd. on 50).

Chen, A. (2015). The ever-growing ed-tech market. *The Atlantic* (Ctd. on 133).

Chopra, R. (2013). Student debt swells, federal loans now top a trillion. *Consumer Financial Protection Bureau* (Ctd. on 133).

Colorado Commission on Higher Education. (2010). *Accelerated erosion of state support—fifty percent overall reduction: What it means to Colorado public higher education.* https://web.archive.org/web/20170203215015/http://highered.colorado.gov/Finance/Budget/2010/201011_HEStateSupport.pdf. (Ctd. on 135)

Colosi, P. J. (2020). Christian personalism versus utilitarianism: An analysis of their approaches to love and suffering. *The Linacre Quarterly, 87*(4), 425–37 (Ctd. on 146).

Committee on CCCC Language. (1974). Students' right to their own language. *College Composition and Communication, 25*(3), 1–18. doi: 10.2307/356219 (Ctd. on 56, 59).

Cook, S. (2020). US schools leaked 24.5 million records in 1,327 data breaches since 2005 [blog post]. *Comparitech.* https://www.comparitech.com/blog/vpn-privacy/us-schools-data-breaches/ (Ctd. on 167)

Copenhaver, A. (2019). Violent video games as scapegoat after school shootings in the United States. In G. A. Crews (Editor), *Handbook of research on mass shootings and multiple victim violence* (Pages 243–66). IGI Global. (Ctd. on 147).

Cottom, T. M. (2014). The new old labor crisis. *Slate* (Ctd. on 96).

Crawford, S. P. (2011, December 3). The new digital divide. *The New York Times.* (Ctd. on 125).

Csiszar, A., Gingras, Y., Power, M., Wouters, P., Griesemer, J. R., Kehm, B. M., de Rijcke, S., Stöckelová, T., Fanelli, D., Sismondo, S., et al. (2020). *Gaming the metrics: Misconduct and manipulation in academic research.* MIT Press. (Ctd. on 144).

Cushman, E. (2013). Wampum, sequoyan, and story: Decolonizing the digital archive. *College English, 76*(2), 115–135 (Ctd. on 136).

Dadniel. (2007, October 21). *George Carlin: Saving the planet* [Video]. YouTube. https://youtu.be/7W33HRc1A6c. (Ctd. on 148)

Davidson, C. N. (2011). *Now you see it: How the brain science of attention will transform the way we live, work, and learn.* Viking New York, NY. (Ctd. on 151).

Davidson, C. N. (2017a). *The new education: How to revolutionize the university to prepare students for a world in flux.* Hachette UK. (Ctd. on 19, 20).

Davidson, C. N. (2017b). Why picking a major is a bad idea for college kids. *Time* (Ctd. on 26).

Davis, J. (2013). A radical way of unleashing a generation of geniuses. *Wired* (Ctd. on 72).

DePew, K. E., & Lettner-Rust, H. (2009). Mediating power: Distance learning interfaces, classroom epistemology, and the gaze. *Computers and Composition*, 26(3), 174–189 (Ctd. on 136).

DeRosa, R. (2018). Student-created open "textbooks" as course communities. *Open Pedagogy Notebook* (Ctd. on 26).

Deterding, S. (2014). Eudaimonic design, or: Six invitations to rehtink gamification. In M. Fuchs, S. Fizek, P. Ruffino, & N. Schrape (Editors), *Rethinking gamification* (Pages 305–23). Meson Press. (Ctd. on 144).

DeVoss, D. N., Cushman, E., & Grabill, J. T. (2005). Infrastructure and composing: The when of new-media writing. *College composition and communication*, 14–44 (Ctd. on 136).

Dewey, J. (2007). *Experience and education*. Simon & Schuster. (Ctd. on 56, 59).

Dominus, S. (2013). How to get a job with a philosophy degree. *The New York Times Magazine* (Ctd. on 145).

Doyle, T. (2012). *Learner-centered teaching: Putting the research on learning into practice*. Stylus Publishing, LLC. (Ctd. on 76).

Dreher, R. (2017). The trouble with trump. *The American Conservative* (Ctd. on 39).

Dweck, C. S. (2008). *Mindset: The new psychology of success*. Random House Digital, Inc. (Ctd. on 31).

Eli Review. (2016). Jeff Grabill's 2016 Computers & Writing keynote address – video, transcript, reactions. *The Eli Review Blog*. https://elireview.com/2016/05/24/grabill-cw-keynote/ (Ctd. on 139)

El-Taraboulsi, S. (2011). Spaces of citizenship: Youth civic engagement and pathways to the January 25 revolution. In S. Stroud & B. Ibrahim (Editors), *Youth activism and public space in Egypt* (Pages 10–19). John D. Gerhart Center for Philanthropy; Civic Engagement. (Ctd. on 124).

Eyler, J. R. (2018). *How humans learn: The science and stories behind effective college teaching*. West Virginia University Press. (Ctd. on ix).

Fanon, F. (2008). *Black skin, white masks* (R. Philcox, Translator). Grove Press. (Ctd. on 154).

Farber, J. (1968). *The student as nigger*. A Few London Libertarians. (Ctd. on 107).

Fedor, C., Noon, E., & Sines, T. (2019). Final project: Refugees and citizenship. *Introduction to Digital Publishing*. http://stevensonenglish.org/eng256-om1s-licastro19/2019/05/12/final-project-refugees-and-citizenship/ (Ctd. on 103)

Feldstein, M. (2013). What faculty should know about adaptive learning. *eLiterate*. https://eliterate.us/faculty-know-adaptive-learning/ (Ctd. on 15)

Fitzpatrick, K. (2019). *Generous thinking: A radical approach to saving the university*. JHU Press. (Ctd. on 23).

Fixsen, A. (2017). *Feeling our way: An ethnographic exploration of university staff experiences of 'soft skills' learning and development programmes* (Doctoral dissertation). University of Westminster. (Ctd. on 145).

Flaherty, C. (2014). 'Critical' organizing. *Inside Higher Ed*. https://www.insidehighered.com/news/2014/03/25/union-event-focuses-adjuncts-role-changing-their-own-working-conditions (Ctd. on 95)

Flaherty, C. (2020). Diversity work, interrupted. *Inside Higher Ed*. https://www.insidehighered.com/news/2020/10/07/colleges-cancel-diversity-programs-response-trump-order (Ctd. on 106)

Franklin, U. (1999). *The real world of technology*. House of Anansi. (Ctd. on 164).

Fred Rogers. (2020, December 1). In *Wikipedia*. https://en.wikipedia.org/wiki/Fred_Rogers. (Ctd. on 111)

Freire, P. (2014). *Pedagogy of the oppressed* (30th anniversary edition). Bloomsbury. (Ctd. on xvii, 4, 56, 65, 71, 123, 152).

Friend, C. (2015). Cfp: The purpose of education. *Hybrid Pedagogy*. https://hybridpedagogy.org/cfp-the-purpose-of-education (Ctd. on 133–135)

Friend, C., & Gilliard, C. (2018). Platforms [podcast episode]. *HybridPod*. https://hybridpedagogy.org/platforms (Ctd. on 131)

Friend, C., & Morris, S. M. (2013). Listening for student voices. *Hybrid Pedagogy*. https://hybridpedagogy.org/listening-for-student-voices (Ctd. on 37)

Friend, C., Shaffer, K., Inoue, A. B., & Bessette, L. S. (2015). Assessment and generosity [podcast episode]. *HybridPod*. https://hybridpedagogy.org/assessment (Ctd. on 76)

Fukuyama, F. (1989). The end of history? *The National Interest*, (16), 3–18 (Ctd. on 148).

Gessen, M. (2016). Autocracy: Rules for survival. *The New York Review of Books* (Ctd. on 39).

Gilliam, T. (Director). (1985). *Brazil* [Film]. Twentieth Century Fox. (Ctd. on 153).

Giroux, H. A. (2001). *Theory and resistance in education: Towards a pedagogy for the opposition*. Greenwood Publishing Group. (Ctd. on 148).

Glover, D. (2018, May 5). *Childish Gambino - this is America (official video)* [Video]. YouTube. https://youtu.be/VYOjWnS4cMY. (Ctd. on 102)

Godwin, M. (1994). Meme, counter-meme. *Wired*. https://www.wired.com/1994/10/godwin-if-2/ (Ctd. on 170, 171)

Gold, J. (2017). Learning how to know in 2017. *Teaching Tolerance* (Ctd. on 39).

González, N., Moll, L. C., & Amanti, C. (2006). *Funds of knowledge: Theorizing practices in households, communities, and classrooms*. Routledge. (Ctd. on 59).

Goodin, D. (2017). Now sites can fingerprint you online even when you use multiple browsers. *Ars Technica* (Ctd. on 129).

Gould, S. J. (1996). *The mismeasure of man*. WW Norton. (Ctd. on 165).

Gould, S. J., & Vrba, E. S. (1982). Exaptation—A missing term in the science of form. *Paleobiology*, 4–15 (Ctd. on 146).

Grabill, J. T. (2003). On divides and interfaces: Access, class, and computers. *Computers and composition*, 20(4), 455–472 (Ctd. on 136).

Graff, D. (2016). An update to our AdWords policy on lending products. *The Keyword* (Ctd. on 131).

Groom, J. (2014). How the Web was ghettoized for teaching and learning in higher ed? *Bava Tuesdays* (Ctd. on 48).

Hamilton. (2017, June 28). *The Hamilton mixtape: Immigrants (we get the job done)* [Video]. YouTube. https://youtu.be/6_35a7sn6ds. (Ctd. on 102)

Hanstedt, P. (2018). *Creating wicked students: Designing courses for a complex world*. Stylus Publishing, LLC. (Ctd. on 19).

Harasim, L. (2017). *Learning theory and online technologies*. Taylor & Francis. (Ctd. on 104).

Haraway, D. (1991). A cyborg manifesto: Science, technology, and socialist-feminism in the late twentieth century. *Simians, cyborgs, and women: The reinvention of nature* (Pages 149–81). Routledge. (Ctd. on 165).

Hartley, D. (2007). Personalisation: The emerging 'revised' code of education? *Oxford Review of Education*, 33(5), 629–642. https://doi.org/10.1080/03054980701476311 (Ctd. on 14)

Harvey, D. (2003). The fetish of technology: Causes and consequences. *Macalester International*, 13(1), 7 (Ctd. on 8).

Hasinoff, A. (2017a). Experiments in self-assessment. *Thinq Studio*. https://web.archive.org/web/20181002035300/http://thinq.studio/digped-2017/experiments-in-self-assessment/ (Ctd. on 76)

Hasinoff, A. (2017b). Getting away from grades: Student self-assessment. *Thinq Studio*. https://web.archive.org/web/20181002035303/http://thinq.studio/digped-2017/self-assessment/ (Ctd. on 76)

Heidebrink-Bruno, A. (2014). Cracking open the curriculum. *Hybrid Pedagogy* (Ctd. on 26).

Hepp, A., & Krotz, F. (2014). *Mediatized worlds: Culture and society in a media age*. Springer. (Ctd. on 7).

Hill, D., & Kumar, R. (2012). *Global neoliberalism and education and its consequences*. Routledge. (Ctd. on 144).

Hill, P. (2013). University of Phoenix patents adaptive activity stream for its learning platform. *eLiterate*. https://eliterate.us/university-of-phoenix-patents-adaptive-activity-stream-for-its-learning-platform/ (Ctd. on 15)

Hobson, T. (2011). Loving them just as they are. *Teacher Tom: Teaching and Learning from Preschoolers* (Ctd. on 33).

hooks, b. (2003). *Teaching community: A pedagogy of hope*. Routledge. (Ctd. on 118).

hooks, b. (2014). *Teaching to transgress*. Routledge. (Ctd. on 56, 60, 76, 78).

House Committee on Education & the Workforce Democratic Staff. (2014). *The just-in-time professor: A staff report summarizing eforum responses on the working conditions of contingent faculty in higher education*. U.S. House of Representatives. https://web.archive.org/web/20141215234740/http://democrats.edworkforce.house.gov/sites/democrats.edworkforce.house.gov/files/documents/1.24.14-AdjunctEforumReport.pdf. (Ctd. on 95)

Howard, R. M. (1999). *Standing in the shadow of giants: Plagiarists, authors, collaborators*. Greenwood. (Ctd. on 139).

Howard, R. M. (2007). Understanding "Internet plagiarism". *Computers and Composition*, 24(1), 3–15. https://doi.org/10.1016/j.compcom.2006.12.005 (Ctd. on 139)

Hudson, E. (2015). Teaching as wayfinding. *Hybrid Pedagogy*. https://hybridpedagogy.org/teaching-as-wayfinding/ (Ctd. on 42)

Inoue, A. B. (2019a). Assessing english so that people stop killing each other. *Labor-based grading contracts: Building equity and inclusion in the compassionate writing classroom* (Pages 305–11). WAC Clearinghouse. (Ctd. on 107).

Inoue, A. B. (2019b). Problematizing grading and the white *habitus* of the writing classroom. *Labor-based grading contracts: Building equity and inclusion in the compassionate writing classroom* (Pages 21–48). WAC Clearinghouse. (Ctd. on 111).

International Military Tribunal. (1947). Second day, Wednesday, 11/21/1945, part 04. *Trial of the major war criminals before the International Military Tribunal* (Pages 98–102). IMT. http://law2.umkc.edu/faculty/projects/ftrials/nuremberg/jackson.html. (Ctd. on 171)

Jhangiani, R. (2017). Why have students answer questions when they can write them? [personal blog]. *That Psych Prof* (Ctd. on 26).

Jorre de St Jorre, T., & Oliver, B. (2018). Want students to engage? contextualise graduate learning outcomes and assess for employability. *Higher Education Research & Development, 37*(1), 44–57 (Ctd. on 145).

Kahn, G. (2014). College in a box: Textbook giants are now teaching classes. *Slate* (Ctd. on 37).

Kamenetz, A. (2014, July 8). The collapse of corinthian colleges [radio broadcast]. On *Morning Edition*. NPR (Ctd. on 14).

Kendzior, S. (2014). The minimum wage worker strikes back. *Medium* (Ctd. on 82).

Kinloch, V. F. (2005). Revisiting the promise of" students' right to their own language": Pedagogical strategies. *College Composition and Communication, 83*–113 (Ctd. on 59).

Kohn, A. (2011). The case against grades. *Educational Leadership.* https://www.alfiekohn.org/article/case-grades/ (Ctd. on 76)

Kolowich, S. (2013). The new intelligence. *Inside Higher Ed.* https://www.insidehighered.com/news/2013/01/25/arizona-st-and-knewtons-grand-experiment-adaptive-learning (Ctd. on 15)

Korzybski, A. (1958). *Science and sanity: An introduction to non-Aristotelian systems and general semantics.* Institute of GS. (Ctd. on 145).

Kramer, J., Dong, Y., & Wolfsheimer, K. (2019). Neither hair nor their podcast. *Introduction to Digital Publishing.* http://stevensonenglish.org/eng256-on1s-licastro19/2019/05/11/neither-hair-nor-their-podcast/ (Ctd. on 103)

Lawrence, S. D. (2012). Turnitin, ProQuest to add dissertation database. *Education News | Technology* (Ctd. on 141).

Lee, J. (2014). Exactly how often do police shoot unarmed Black men? *Mother Jones* (Ctd. on 82).

Lenhart, A., Duggan, M., Perrin, A., Stepler, R., Rainie, H., Parker, K., et al. (2015). *Teens, social media & technology overview 2015.* Pew Research Center [Internet & American Life Project]. (Ctd. on 9).

Lepore, J. (2009). Not so fast. *The New Yorker*. https://www.newyorker.com/magazine/2009/10/12/not-so-fast (Ctd. on 17)

Lessig, L. (2004). *Free culture: How big media uses technology and the law to lock down culture and control creativity*. Penguin. (Ctd. on 138).

Lessig, L. (2007, March). *Larry Lessig says the law is strangling creativity* [Video]. TED. (Ctd. on 137).

Lewis, C. S. (2002). *On stories: And other essays on literature*. Houghton Mifflin Harcourt. (Ctd. on 147).

Licastro, A. (2017). Learning at the intersections. *Hybrid Pedagogy*. https://hybridpedagogy.org/learning-intersections/ (Ctd. on 99)

Linden, M. (2017). Trump's America and the rise of the authoritarian personality. *The Conversation* (Ctd. on 79).

Lorde, A. (2018). *The master's tools will never dismantle the master's house*. Penguin UK. (Ctd. on 145).

Lorde, A. (2020). *The cancer journals*. Penguin Classics. (Ctd. on 81, 83, 84).

Losh, E. (2012). Hacktivism and the humanities: Programming protest in the era of the digital university. In M. K. Gold (Editor), *Debates in the digital humanities* (Pages 161–186). University of Minnesota Press. (Ctd. on 14).

Macgillivray, A. (2009). Some questions related to Google News and the Associated Press [blog post]. *Google Public Policy Blog* (Ctd. on 141).

Madden, M., Lenhart, A., & Fontaine, C. (2017). *How youth navigate the news landscape*. John S. and James L. Knight Foundation. (Ctd. on 131).

Maggio, J. (Writer), & Smith, M. (Writer). (2010, May 4). College, Inc. (Season 28, Episode 9) [TV series episode]. In C. Durrance, J. Maggio, & M. Smith (Executive Producers), *Frontline*. PBS. (Ctd. on 13).

Mason, J. (2002). *Researching your own practice: The discipline of noticing*. Psychology Press. (Ctd. on 40).

Matsuda, P. K. (2006). The myth of linguistic homogeneity in U.S. college composition. *College English, 68*(6), 637–51 (Ctd. on 57).

Mattei, U. (2014). Future generations now! A commons-based analysis. In U. M. Gilda Farrell Saki Bailey (Editor), *Protecting future generations through commons* (Pages 9–26). Council of Europe. (Ctd. on 147).

McCalmont, L. (2015). Walker urges professors to work harder. *Politico* (Ctd. on 135).

McCarthy, C. (2014). Adjunct professors fight for crumbs on campus. *The Washington Post* (Ctd. on 96).

McGonigal, J. (2011). *Reality is broken: Why games make us better and how they can change the world*. Penguin. (Ctd. on 151).

Mehta, J. (2013). Why American education fails: And how lessons from abroad could improve it. *Foreign Affairs, 92*, 105 (Ctd. on 135).

Menebrea, L. F. (1837). Sketch of the analytical engine invented by Charles Babbage Esq. (A. Lovelace, Translator). In A. Henfrey, T. H. Huxley, & R. Taylor (Editors), *Scientific memoirs: Selected from the transactions of foreign academics of science and from foreign journals* (Page 696). Taylor. (Ctd. on 168).

Mercy Corps. (2012). *Civic engagement of youth in the Middle East and North Africa: An analysis of key drivers and outcomes.* Mercy Corps. (Ctd. on 124).

Millozzi, V. (2015). From E.U. policies to the economic crisis: The rise of the neoliberalagenda and the demise of public education in Italy. In G. Grollios, A. Liambas, & P. Pavlidis (Editors), *Critical education in the era of crisis* (Pages 490–97). Aristotle University of Tessaloniki Greece Faculty of Education. (Ctd. on 149).

MIT Libraries. (2020, June 2). *Elsevier fact sheet.* https://libraries.mit.edu/scholarly/publishing/elsevier-fact-sheet/. (Ctd. on 138)

Morini, L. (2016). Play as bulwark of uselessness. *First Person Scholar.* http://www.firstpersonscholar.com/play-as-bulwark-of-uselessness/ (Ctd. on 143)

Morozov, E. (2013). *To save everything, click here: The folly of technological solutionism.* Public Affairs. (Ctd. on 8).

Morris, S. M. (2014). We may need to amputate: MOOCs, resistance, #futureed. *Hybrid Pedagogy* (Ctd. on 10).

Morris, S. M. (2018). Imagination as a precision tool for change [blog post]. *Sean Michael Morris* (Ctd. on 79).

Morris, S. M., & Stommel, J. (2012). Hacking the screwdriver: Instructure's canvas and the future of the lms. *Hybrid Pedagogy* (Ctd. on 136).

Morris, S. M., & Stommel, J. (2018). *An urgency of teachers: The work of critical digital pedagogy.* Hybrid Pedagogy Inc. (Ctd. on xiv).

Moyle, A. (Director). (1990). *Pump up the volume* [Film]. New Line Cinema. (Ctd. on 126).

National Council of Teachers of English. (1974, November 30). Resolution on the students' right to their own language. https://ncte.org/statement/righttoownlanguage/. (Ctd. on 57)

Nelson, T. H. (1987). *Computer lib/Dream machines.* Microsoft Press. (Ctd. on 129).

Neumann, J. W. (2013). Advocating for a more effective critical pedagogy by examining structural obstacles to critical educational reform. *The Urban Review, 45*(5), 728–740 (Ctd. on 14).

Noenickx, C. (2019, July 28). 'Where I'm from': A crowdsourced poem that collects your memories of home [radio broadcast]. On *Morning Edition*. NPR (Ctd. on 101).

North American Montessori Center. (2008). Montessori philosophy: Praise vs. encouragement [blog post]. *NAMC Montessori Teacher Training Blog* (Ctd. on 30).

Obama, B. (2020). *A promised land*. Crown. (Ctd. on xvi–xviii).

Office of Research, Evaluation, and Statistics. (1997). Unemployment insurance. *Social Security Programs in the United States*. https://web.archive.org/web/20041118031652/http://www.ssa.gov/policy/docs/progdesc/sspus/unemploy.pdf (Ctd. on 96)

Oliver, N. C. (2014). I just don't want to die an adjunct [blog post]. *Nathaniel C. Oliver*. https://web.archive.org/web/20140827022447/http://nathanielcoliver.com/2014/08/22/i-just-dont-want-to-die-an-adjunct (Ctd. on 97)

Olkon, S. (2009). These dorms major in luxury. *Chicago Tribune* (Ctd. on 88).

O'Reilley, M. R. (1989). "Exterminate…the brutes"—and other things that go wrong in student-centered teaching. *College English, 51*(2), 142–146 (Ctd. on 107, 108, 110).

O'Reilley, M. R. (1998). *Radical presence: Teaching as contemplative practice*. Boynton/Cook. (Ctd. on 110).

Owens, T., & Mir, R. (2012). If (!isnative()){return false;}: De-people-ing native peoples in Sid Meier's *Colonization*. *Play the Past*. http://www.playthepast.org/?p=2509 (Ctd. on 154)

Papert, S. (1993). *The children's machine: Rethinking school in the age of the computer*. Basic Books. (Ctd. on xi).

Parker-Rees, R. (1999). Protecting playfulness. In H. Moylett & L. Abbott (Editors), *Early education transformed* (Pages 61–72). Psychology Press. (Ctd. on 148).

Parrish, S. (2019). The museum as medium: An experiment in interdisciplinary project-based learning [personal blog]. *Sarah Parrish* (Ctd. on 23).

Patton, S. (2014). The MLA conference and the curious case of the $6.25 granola bar. *Community*. https://community.chronicle.com/news/273-the-mla-conference-and-the-curious-case-of-the-6-25-granola-bar (Ctd. on 86)

Patton, S. (2015). Dear student: No, I won't change the grade you deserve. *Chronicle Vitae*, 13th (Ctd. on 135).

Perry, M., Johnston, W., & Candon, J. (2019). The American dream (choose your own adventure). *Introduction to Digital Publishing*. http://

stevensonenglish.org/eng256-on1s-licastro19/2019/05/10/the-american-dream-choose-your-own-adventure/ (Ctd. on 104)

Perryman-Clarke, S., Kirkland, D., & Jackson, A. (2014). Students' rights to their own language: A critical sourcebook (Ctd. on 59).

Pollan, M. (2015). *The omnivore's dilemma: The secrets behind what you eat.* Listening Library. (Ctd. on 21, 22).

Prendergast, C. (2008). *Buying into english: Language and investment in the new capitalist world* (Volume 31). University of Pittsburgh Pre. (Ctd. on 56, 60).

Price, M. (2002). Beyond" gotcha!": Situating plagiarism in policy and pedagogy. *College Composition and Communication*, 88–115 (Ctd. on 139).

ProQuest. (no date). *Proquest dissertations faq.* http://www.proquest.com/products-services/dissertations/ProQuest-Dissertations-FAQ.html. (Ctd. on 140)

Pryal, K. R. G. (2013). A lecturer's almanac. *Hybrid Pedagogy* (Ctd. on 96).

Reitenauer, V. L. (2017). 'a practice of freedom': Self-grading for liberatory learning. *Radical Teacher, 107*, 60–63 (Ctd. on 76).

Rheingold, H. (2012). *Net smart: How to thrive online.* MIT Press. (Ctd. on 66, 125).

Rich, A. (1977). Claiming our education. *The Common Woman, 1*(2), 9–11 (Ctd. on 79).

Richtel, M. (2012, May 29). Wasting time is new divide in digital era. *The New York Times.* (Ctd. on 125).

Rogers, F. (1995). *You are special: Neighborly words of wisdom from Mister Rogers.* Penguin Publishing Group. (Ctd. on 151).

Rooney, S., & Rawlinson, M. (2016). Narrowing participation? Contesting the dominant discourse of employability in contemporary higher education. *Journal of the National Institute for Career Education and Counselling, 36*(1), 20–29 (Ctd. on 145).

Rorabaugh, P., Morris, S. M., & Stommel, J. (2013). Beyond rigor. *Hybrid Pedagogy.* https://hybridpedagogy.org/beyond-rigor/ (Ctd. on 151)

Rorabaugh, P., & Stommel, J. (2012). Hybridity, pt. 3: What does Hybrid Pedagogy do? *Hybrid Pedagogy* (Ctd. on 10).

Rushdie, S. (1991). The location of brazil. *Imaginary homelands: Essays and criticism 1981–1991.* Penguin. (Ctd. on 153).

Sacco, J. (2003). *Palestine.* Jonathan Cape. (Ctd. on 152).

Sackstein, S. (2015). *Hacking assessment: 10 ways to go gradeless in a traditional grades school.* Times 10 Publications. (Ctd. on 78).

Said, E. W. (2004). *Humanism and democratic criticism.* Columbia University Press. (Ctd. on 123).

Schinske, J., & Tanner, K. (2014). Teaching more by grading less (or differently). *CBE—Life Sciences Education*, *13*(2), 159–166 (Ctd. on 76, 78).

Schneider, J. (2016). America's not-so-broken education system: Do U.S. schools really need to be disrupted? *The Atlantic* (Ctd. on 135).

Schneider, N. (2014). The real trouble with disruption. *Vice*. https://www.vice.com (Ctd. on 17)

Schuman, R. (2013). Hey search committees who haven't requested interviews yet: F&%# YOU [blog post]. *Pan Kisses Kafka*. https://pankisseskafka.com/2013/12/04/hey-search-committees-who-havent-requested-interviews-yet-fuck-you/ (Ctd. on 87)

Selber, S. (2004). *Multiliteracies for a digital age*. SIU Press. (Ctd. on 135–137).

Selfe, C. L., & Selfe, R. J. (1994). The politics of the interface: Power and its exercise in electronic contact zones. *College composition and communication*, *45*(4), 480–504 (Ctd. on 136).

Shehata, D. (2008). *Youth activism in Egypt* (Arab Reform Brief). Arab Reform Initiative, University of Jordan. (Ctd. on 124).

Shor, I., & Freire, P. (1987). *A pedagogy for liberation: Dialogues on transforming education*. Greenwood Publishing Group. (Ctd. on 60).

Simmons, W. M., & Grabill, J. T. (2007). Toward a civic rhetoric for technologically and scientifically complex places: Invention, performance, and participation. *College Composition and Communication*, 419–448 (Ctd. on 136).

Singer, N. (2013). They loved your GPA then they saw your tweets. *The New York Times* (Ctd. on 131).

Singham, M. (2005). Away from the authoritarian classroom. *Change: The Magazine of Higher Learning*, *37*(3), 50–57. https://doi.org/10.3200/CHNG.37.3.50-57 (Ctd. on 75)

Snowden, E. (2019). *Permanent record*. Macmillan. (Ctd. on xii).

Solomon, B. C., & Ismail, I. (2017, March 24). *The displaced* [Video]. The New York Times. https://www.nytimes.com/video/magazine/100000005005806/the-displaced.html. (Ctd. on 103)

Spelic, S. (2016). Never the same Twitter. *Identity, Education, and Power* (Ctd. on 43).

Spielberg, S. (Director). (1993). *Jurrasic park* [Film]. Amblin Entertainment. (Ctd. on 146).

Star, K. (2013). Doing useful work using games. *International Conference on Games and Learning Alliance*, 316–23 (Ctd. on 147).

Stenros, J. (2015). *Playfulness, play, and games: A constructionist ludology approach* (Doctoral dissertation). Tampere University. (Ctd. on 147).

Stommel, J. (2014). Critical digital pedagogy: A definition. *Hybrid Pedagogy* (Ctd. on 21, 28, 56, 60).

Stommel, J. (2015). Who controls your dissertation? *Chronicle of Higher Education*, 7 (Ctd. on 139).

Stommel, J. (2017). Why i don't grade [blog post]. *Jesse Stommel.* https://www.jessestommel.com/why-i-dont-grade/ (Ctd. on 76)

Stommel, J., Friend, C., & Morris, S. M. (2020a). Critical digital pedagogy: A collection. *Hybrid Pedagogy.* https://hybridpedagogy.org/critical-digital-pedagogy/ (Ctd. on xiii)

Stommel, J., Friend, C., & Morris, S. M. (Editors). (2020b). *Critical digital pedagogy: A collection.* Hybrid Pedagogy Inc. https://cdpcollection.pressbooks.com. (Ctd. on xiv)

Strauss, V. (2014). Why school isn't for children anymore—teacher [blog post]. *Answer Sheet.* http://www.washingtonpost.com/blogs/answer-sheet/wp/2014/03/03/why-school-isnt-for-children-anymore-teacher (Ctd. on 37)

Suits, B. (2014). *The grasshopper: Games, life and utopia.* Broadview Press. (Ctd. on 148).

Tannock, S. (2017). No grades in higher education now! revisiting the place of graded assessment in the reimagination of the public university. *Studies in Higher Education*, 42(8), 1345–1357 (Ctd. on 76, 79).

Thatcher, M. (1980, June 25). *Press conference for American correspondents in London.* Margaret Thatcher Foundation. https://www.margaretthatcher.org/document/104389. (Ctd. on 149)

The Emmy Amards. (2008, March 26). *Fred Rogers acceptance speech - 1997* [Video]. YouTube. https://youtu.be/Upm9LnuCBUM. (Ctd. on 111)

The Fair Housing Center of Greater Boston. (n.d.). 1934–1968: FHA mortgage insurance requirements utilize redlining. *Historical Shift from Explicit to Implicit Policies Affecting Housing Segregation in Eastern Massachusetts* (Ctd. on 130).

The Guardian. (2017, July 5). *Limbo: A virtual experience of waiting for asylum* [Video]. YouTube. https://youtu.be/AyWLvrWBKHA. (Ctd. on 103)

The Highlander Editorial Board. (2014). Editorial: Standardized testing has created standardized students with useless skills. *The Highlander* (Ctd. on 37).

Thompson, D. (2020). We're never going back to the 1950s. *The Atlantic.* https://www.theatlantic.com/ideas/archive/2020/12/how-2020-shattered-shared-reality/617398/ (Ctd. on xii)

Thoreau, H. D. (1995). *Walden; or, life in the woods.* Project Gutenberg. https://www.gutenberg.org/files/205. (Ctd. on 7)

Tufekci, Z. (2014). What happens to #ferguson affects Ferguson: Net neutrality, algorithmic filtering and Ferguson. *The Message.* https://medium.com/message (Ctd. on 16)

Tyack, D., & Tobin, W. (1994). The "grammar" of schooling: Why has it been so hard to change? *American Educational Research Journal, 31*(3), 453–479 (Ctd. on 17).

U.N. Human Development Report Office. (2013). *Public expenditure on education* (Human Development Reports). United Nations Development Programme. http://hdr.undp.org/en/content/expenditure-education-public-gdp. (Ctd. on 133)

Vanderhye v. iParadigms LLC. http://caselaw.findlaw.com/us-4th-circuit/1248473.html (Ctd. on 141)

Vancouver Community College Faculty Association. (2013, May). *A program for change: Real transformation over two decades.* https://web.archive.org/web/20150527071603/http://vccfa.ca/newsite/wp-content/uploads/2012/05/Access-the-Program-for-Change-May-2013.pdf. (Ctd. on 96)

van Manen, M. (2016). *The tone of teaching: The language of pedagogy.* Routledge. (Ctd. on 34).

Vie, S. (2013). Turn it down, don't turnitin: Resisting plagiarism detection service by talking about plagiarism rhetorically. *Computers and Composition Online* (Ctd. on 139).

Walls, D. M., Schopieray, S., & DeVoss, D. N. (2009). Hacking spaces: Place as interface. *Computers and Composition, 26*(4), 269–287 (Ctd. on 136).

Walsh, J. (2018). Higher education in the twenty-first century. *Higher education in Ireland, 1922–2016: Politics, policy and power—a history of higher education in the Irish state* (Pages 483–94). Palgrave Macmillan. (Ctd. on 144).

Walsh, L. (2013). *Scientists as prophets: A rhetorical genealogy.* Oxford University Press. (Ctd. on 14–16).

Watson, J. (2012). *Reality ends here: Environmental game design and participatory spectacle* (Doctoral dissertation). University of Southern California. (Ctd. on 147).

Watson, J. (2013). Gamification: Don't say it, don't do it, just stop. *Media Commons, September, 21* (Ctd. on 144).

Watters, A. (2020). Educational crises and ed-tech: A history. *Hack Education* (Ctd. on 5).

Wesch, M. (2015a, August 26). *The sleeper* [Video]. YouTube. https://youtu.be/mZedcQoYoiw. (Ctd. on 9)

Wesch, M. (2015b, August 25). *What Baby George taught me about learning* [Video]. YouTube. https://youtu.be/hbRFAq9XEV0. (Ctd. on 9)

Westby, E. L., & Dawson, V. (1995). Creativity: Asset or burden in the classroom? *Creativity Research Journal*, *8*(1), 1–10 (Ctd. on 72).

Westheimer, J., & Kahne, J. (1998). Education for action: Preparing youth for participatory democracy. In W. Ayers, J. A. Hunt, & T. Quinn (Editors), *Teaching for social justice: A democracy and education reader* (Pages 1–20). New Press. (Ctd. on 124).

Wilson, R. A. (2012). *The Illuminati papers*. Ronin Publishing. (Ctd. on 145).

World Meteorological Association. (2017). Asperitas. *International Cloud Atlas* (Ctd. on 50).

Wysocki, A. F., & Jasken, J. I. (2004). What should be an unforgettable face... *Computers and Composition*, *21*(1), 29–48 (Ctd. on 136).

Zernike, K. (2016). A sea of charter schools in Detroit leaves students adrift. *The New York Times* (Ctd. on 135).

Zuboff, S. (2015). Big other: Surveillance capitalism and the prospects of an information civilization. *Journal of Information Technology*, *30*(1), 75–89 (Ctd. on 129).

Index

A

Action 84
Active learning 32
Activism 124
Adjuncts 95
. . . . see also Labor
Administration 160
Advocacy 59, 66, 122
Affirmative action 156
Agency . 15, 58, 59, 66, 67, 122, 133, 135–137, 139, 141, 154
al-Sisi, Abdel Fattah 123
Algorithms 15, 46, 130, 163
Alternative facts 46
Altruism 100
Anti-racist practices 108
Arbery, Ahmaud 117
Assessment . 88, 108, 109, 111, 112, 135
. . . . see also Grading
Asylum seekers 100, 101, 104
Attention 37, 110–112
Authority . 72, 75–77, 105, 121, 134
Automation 139, 168
Autonomy 70, 133, 134, 136

B

Babbage, Charles 168
Banking model . . . 4, 31, 32, 37, 126
. . . . see also Critical Pedagogy
Becoming 111, 113
Behaviorism 4, 71
Bias 22, 52, 54, 58, 60, 101, 102, 119
Big data 16, 166–168
Blackboard . . . . see LMS
Blackness . . . 82, 108, 117, 168, 172

Blogs 49
Borders 136
Brazil 153
Brooks, Rayshard 117
Brown, Michael 82, 83
Building 51
. . . . see also Creativity
Bureaucracy 88

C

Canvas . . . . see LMS
Capitalism 129, 132, 147
Care 31, 88, 118
Citizenship 99–101, 121, 122
Civic engagement 124, 135
Civics 45
Civilization (game) 154–155
Clouds, meteorological 50
Collaboration 19
Collaborativism 104
Colonialism 148, 155
Comic books 152
Compassion 111, 113
Conformity 69, 70, 72
Consent 131–132
Constructivism 71, 104
Contemplative practices 110
Content delivery 119, 126
. . . . . . see also Banking model, Critical pedagogy
Contingency . . . . see Labor
Contracts 95
. . . . see also Tenure
Control 75, 122, 129, 134, 153, 165–167, 170
Copyright 138, 141
Coronavirus . . . . see COVID-19
Costs

hidden 139
of technology 126
of travel 86
systemic 135
travel 87
Costs of education systems . . . 135
COVID-19 . . . . 3, 118
Creativity . . . 91, 121, 126, 152, 153
Critical pedagogy 9, 32, 152
Critical thinking 121, 122

D

D2L . . . . *see* LMS
Davis, Jordan 83
Debt, student loan 133, 135
Democracy 135, 138
Dialogue 113, 124
Digital citizenship 45
Digital divide 125, 127, 130
Digital fluency 9, 45, 53
Digital Pedagogy Lab 7
Digital redlining 130–131
Dinosaurs 146
Discussion boards . 46, 48, 101, 170
Disruption 17
Diversity 106, 109, 120
Domain names 50
Domain of One's Own 48–53
Dominant Discourse 108
Dorms, extravagance of 88

E

Early childhood education . . 30–34
Echo chambers 46
Ed-tech 83, 135, 161, 163–173
market value 133
Egypt 121–124
el-Sisi, Abdel Fattah . . . . . *see* al-Sisi

Election, U.S. presidential . . 40, 41,
46, 99, 164, 167
Eli Review 137–139
Emotions 29, 32
Empathy . . 100, 101, 104–106, 123
. . *see also* Rhetorical empathy
Employability 145
Empowerment 38, 49
Encouragement 30
Epistemology 21, 123, 145
Equity 135, 165
Escapism 147
Ethos 14
Evaluation 30
 . . . . *see also* Assessment
course 27, 105

F

Failure 32–33, 52, 92, 120, 126, 127
Fake news 46
Fear 82
Feedback 76, 137
Feminist technologies 165
Ferguson 16, 83
FERPA 167
Filter bubbles 42, 101, 129, 131
Filters 43
First-Year Composition . . . . *see* FYC
Flipped classroom 119
Floyd, George 117
For-profit institutions 13
Free (as in beer) 138
. . . . *see also* Costs
Freire, Paulo 123, 152
Fun 41
Funding 88, 159
venture capitalism 160
FYC 19, 112

G

Games . 56, 103, 104, 126, 143–147, 151, 153, 154, 156, 157
Gamification 144, 147
Garner, Eric 82
Gender . . 60, 83, 145, 166, 169, 172, 173
Generosity 66, 118
Godwin's Law 170
Governance *see* Shared governance
Grading ... 30, 75–76, 78, 108, 119, 120
.... *see also* Assessment
Granola bars 86

H

Habitus 108, 109, 113
Happiness 90, 92
Heutagogy 30
Hope 163
Human condition 110
Humanity 119

I

Identity 120
  categorized 171
  digital 45
  performed 43
Immigrants 100, 101, 104
Immigration policies 108
Imperialism 155
Imperialist white supremacist capitalist patriarchy 119, 120
Independent learning ... 32, 35, 70, 113
Inequality 5, 21, 99, 145
Information overload 39

Intellectual property 134, 141
.... *see also* Turnitin of students 139
Intelligence 165–167, 170
Interdisciplinarity 19–28, 118
Interfaces 47, 49, 67, 131, 136–142, 160
Intimacy 31
Iraq war (invasion by U.S.) 155

J

Jefferson, Atatiana 117
Job market, academic ... 85, 86, 90

K

K-12 34

L

Labor 108, 110, 141
  academic 20, 159
  contingent 86, 88, 97, 135
  contract 160
  exploited 95, 96
Labor gap 85
Labor unions .... *see* Unions
Labor-based grading contracts 108–113
Language 100, 102, 104
Latte 86
Learning management system . *see* LMS, 8
Lecture 33, 35, 70
Lesson plan 36
LGBTQ+ 117, 168
Liberation 123, 147, 152
  ironic 155, 156
Listening 35, 36, 38, 110, 120
Literacies 109

Living wage 85, 87, 90
.... see also Poverty
LMS 5, 16, 46–47, 68, 134, 136, 161, 166
Love 111, 112, 119
Loving 111, 113
Loyalty 87

M

Martin, Trayvon 83
Massive open online course .... see MOOC
McBride, Renisha 83
Meaning-making 40–43
Media . 7, 39, 83, 101, 121, 145, 147
    conglomerates 51
    consumption 41
    sensationalist 122
Military industrial complex ... 164
Mindset 31
MLA 86, 87
MOOC 65, 68, 160
Morsi, Mohamed 122
Mubarak, Hosni 121

N

Naming 50
Nazi use of data 169–171
Neoliberalism 144, 147, 163
Networks 165
News 83
Nostalgia 132
Noticing 40

O

Obsolescence 159
Open access 14
Open education 8, 34, 45

Open-source applications 48
Oppression 119, 129, 152, 155
Outcomes 120, 147
Ownership 134, 135

P

Pandemic .... see COVID-19
Parenting 29
Pattern-seeking 37, 40–43
Personalized learning 14, 83
Platforms 129–132
Play .. 126, 127, 143–149, 151–158
Police brutality 108, 117
Police violence 82, 83
Political ideology
    through gameplay 156
Post-truth politics 46
Poverty 83, 96
.... see also Living wage
Power 33, 60, 76, 108, 136
...... see also Empowerment
Praise 30
.... see also Feedback
Praxis 8–10
Presence 112
Primary education .... see K-12
Privacy 16, 67, 167–168
Privilege 84, 99
Problem-posing education .. 72, 73
.... see also Critical pedagogy
Productivity 145
Profiling 131
Profit 8, 17, 134, 135, 141, 170
ProQuest 67, 134, 139, 140
Public scholarship 10, 14, 34
Publicity 161
Publishing industry 138
Punch cards 168
Purpose of education 124, 148

R

Race 156
    census categorization 171
Racial dominance 108
Racism 82, 83, 117, 118, 173
    institutional 172
Rage 118
Reality 145
Reconciliation 122
Redlining 130
    . . . . . see also Digital redlining
Refugees 100, 101
Research methods 91
Resistance 120
Respect 31, 33, 87
Retention 130
Rhetorical empathy 100–104
Rhetorics
    procedural 154
    racial 154
Rights 117
Rigor 75, 104, 151
Rogers, Fred (aka Mr. Rogers) . 111, 151
Rouse, Fred 117
Rules 75, 119, 126, 154

S

Sabbatical 86
Safety 84
Salary 87
Search committees 87
Search engines 46
Self-awareness 40
Self-directed learning . . . . see Independent learning
Self-esteem 90
Service learning 99, 104
Shared governance 88, 96

Silence 81–84
Social class 127
Social justice 100, 123
Social media . . 9, 10, 40, 46, 51, 65, 66, 82–84, 86, 89, 91, 131
Social mobility 60
Soft skills 145
Speaking 81
Standardization 37, 46, 127
Standards 109
    . . . see also Rigor, Assessment
STEM 123, 146
Student learning outcomes . . . . see Outcomes
Student-centered teaching . . 35, 71, 108, 119
Stunning (as Twitter trend) 41
Subversion . . 69–72, 152, 153, 155, 165
Surveillance 119, 129, 147, 165, 167, 171–173
    capitalism 129
Syllabi 119
Sympathy 100

T

Taylor, Breonna 117
Technological determinism . . . 148
Tenure 85, 96, 127, 161
Tenuresplaining 86
Terrorists 155, 156
Testing
    high-stakes 126
    standardized 166
Tilde spaces 47–48
Tools 8, 9, 49, 51, 160
Tracking 129
Trust 77, 105, 120
Turnitin 67, 134, 139–141
Twitter 82, 126

U

Unemployment 95
Unions 96
Utilitarianism 144, 147
. . . . *see also* Employability

V

Value of education 133
Values 137, 138
Virtual learning environment . . *see* LMS
Virtual-reality simulations 103
Visibility 82
Vulnerability 33, 81, 100

W

Walled garden 48
Wayfinding 42
Ways of knowing *see* Epistemology
WebCT . . . . *see* LMS
Webserver 51
White supremacy . . . 108, 109, 113
Women's rights 108
WordPress 49
World-Wide Web
    2.0 . . . . 9, 129
    as broken architecture . . . 129
    as object of interrogation . . 54
    building for 51
    influence of 45
Writing classes 108

www.ingramcontent.com/pod-product-compliance
Lightning Source LLC
Chambersburg PA
CBHW071621170426
43195CB00038B/1672